HOW TO PASS

STANDARD GRADE

GEOGRAPHY

Dr. Bill Dick

HODDER GIBSON
AN HACHETTE UK COMPANY

The Publishers would like to thank the following for permission to reproduce copyright material:

Acknowledgements
Ordnance Survey Mapping on pages 13, 24 and 76 is reproduced by permission of Ordnance Survey on behalf of HMSO. © Crown copyright 2008. All rights reserved. Ordnance Survey Licence number 100047450.

The Key Ideas are reproduced from the Standard Grade Geography Arrangements document with the permission of the Scottish Qualifications Authority. Questions from past exam papers are also reproduced with the permission of the Scottish Qualifications Authority.

Every effort has been made to trace all copyright holders, but if any have been inadvertently overlooked the Publishers will be pleased to make the necessary arrangements at the first opportunity.

Although every effort has been made to ensure that website addresses are correct at time of going to press, Hodder Gibson cannot be held responsible for the content of any website mentioned in this book. It is sometimes possible to find a relocated web page by typing in the address of the home page for a website in the URL window of your browser.

Hachette's policy is to use papers that are natural, renewable and recyclable products and made from wood grown in sustainable forests. The logging and manufacturing processes are expected to conform to the environmental regulations of the country of origin.

Orders: please contact Bookpoint Ltd, 130 Milton Park, Abingdon, Oxon OX14 4SB. Telephone: (44) 01235 827720. Fax: (44) 01235 400454. Lines are open 9.00–5.00, Monday to Saturday, with a 24-hour message answering service. Visit our website at www.hoddereducation.co.uk. Hodder Gibson can be contacted direct on: Tel: 0141 848 1609; Fax: 0141 889 6315; email: hoddergibson@hodder.co.uk

© Dr. Bill Dick 2005, 2009
First published in 2005 by
Hodder Gibson, an imprint of Hodder Education,
An Hachette UK company
2a Christie Street
Paisley PA1 1NB

This colour edition first published 2009

Impression number   5 4 3
Year   2012 2011

Cover photo © Dennis Flaherty/Riser/Getty Images
Cartoons © Moira Munro 2005, 2008
Illustrations by Tony Wilkins Design
Typeset in 10.5/12pt Frutiger Light by Phoenix Photosetting, Chatham, Kent
Printed in Dubai

A catalogue record for this title is available from the British Library

ISBN-13: 978 0340 973 929

CONTENTS

INTRODUCTION

The Examination

Standard Grade examinations are quite complicated.

The syllabus covers **17 Key Ideas** which are divided into three broad categories, namely, **The Physical Environment, The Human Environment** and **International Issues**.

Questions set across three levels – Foundation (grades 5 and 6), General (grades 3 and 4) and Credit (grades 1 and 2) – are tested across two elements: Knowledge and Understanding, and Enquiry Skills.

There are eight different ways or criteria for testing these elements. You do not have to worry which criteria are being tested in any particular question.

However, those questions dealing with **Knowledge and Understanding** are based on what you know and have learned about various topics.

Those based on **Enquiry Skills** will be based on your opinions and on your ability to use resource material contained in the questions. For example, you may be asked to choose between two or more options and give the reasons for your choice.

Similarly you may be asked to describe the advantages or disadvantages of a given situation, such as a location for an industry, and also to explain your reasons.

These questions are all based on given resources such as maps, diagrams, tables or statements. You are asked to evaluate these resources carefully in your answer.

Mark allocation

The Credit paper has about 80 marks, the General paper about 70 and the Foundation paper about 60 marks. Depending upon how your teacher thinks you have performed in your course, you will be presented either at Foundation/General level or General/Credit level.

Paper times

The times for the three papers are:

Foundation level: 1 hour

General level: 1 hour 25 minutes

Credit level: 2 hours

◆ Not all of the **Key Ideas** are examined at one level.

◆ The examinations are structured so that the Key Ideas are tested across Foundation/General and General/Credit levels.

◆ Therefore you must cover all of these in both your coursework and your revision.

Some tips for revising and sitting the examination

The following tips apply to all of the examinations.

◆ One useful technique when revising individual topics is to use summary note cards on those topics. Make use of past paper questions to test your knowledge or Enquiry Skills. Go over your answers and give yourself a mark for every correct point you make when comparing your answer with your notes. If you work with a classmate, try to mark each other's practice answers.

◆ Practise your diagram drawing skills. Practise your writing skills. Try to ensure that your answers are clearly worded. Try to develop the points which you make in your answer, either by elaborating on points made, or by adding suitable examples to illustrate your points.

◆ Do not give lists, even if you are running out of time. You will lose marks.

◆ If the question asks for an opinion based on a choice, for example, on the suitability of a particular site or area for a development, do not be afraid to refer to negative points such as why the alternatives are not as good. You will get credit for this.

When preparing for the actual examination there are certain good practices which you should follow.

◆ Make sure you have a copy of the examination timetable and have planned a schedule for studying.

◆ Arrive at the examination in plenty of time with the appropriate equipment: pen, pencil, eraser and ruler.

◆ Carefully read the instructions on the paper and those at the beginning of each part of the question.

◆ Answer all of the questions in each paper you sit.

◆ Use the number of marks as a guide to the length of your answer. For example, at Credit level '6 marks' indicates that you need to provide a minimum of six valid points or three developed points.

◆ At General level avoid giving long Credit-style answers to questions. This will save you valuable time in the examination.

◆ Try to include suitable examples in your answer wherever possible. If asked for, provide diagrams, and remember to draw clear, labelled diagrams.

◆ Read the question instructions very carefully. If the question asks you to 'describe', make sure that this is what you do. For example, you may be given a weather map or weather station with symbols and be asked to describe weather in a particular part of Britain. Go through the weather elements shown and describe each of them, referring, for example, to temperature, rainfall, cloud cover, wind speed and direction.

◆ If you are asked to 'explain' you must use phrases such as 'due to' or 'this happens because' or 'this is a result of'. If you describe rather than explain when the instruction asks for explanation, you will lose most of the marks for that question.

◆ If you finish early do not leave the examination room. Use the remaining time to check your answers and go over any questions which you have partially answered, especially Ordnance Survey map questions. By doing this you could perhaps add sufficient marks to your overall total to achieve a much higher grade.

Gathering and Processing Techniques

The Standard Grade syllabus was changed a few years ago, so instead of candidates having to submit an investigation, they now have to answer questions at all levels on 'gathering and processing techniques'. From reports made by the Principal Assessor on examination performance, it is clear that this is one area of the examination which has caused some difficulty for candidates, especially at General and Credit levels. It is important, therefore, to look at this area of the syllabus in some detail.

◆ Questions on these techniques at Foundation level simply require candidates to identify appropriate techniques for gathering and processing information. However, at both General and Credit levels, as well as identifying appropriate techniques and describing them, you also have to justify or give reasons as to why these techniques should be used.

◆ In class you will have gone into some detail about these techniques: where and when certain techniques are appropriate and the main differences in the various ways of gathering and processing data.

◆ The difference between questions at General and Credit levels is that of the complexity of the technique and the data being gathered or processed. The maximum number of marks allotted at General level is 4 marks but at Credit level the majority of these questions are based on 5 or 6 marks.

◆ The part of these questions that candidates seem to have most difficulty with is that which asks candidates to 'justify the use' of various techniques. Although gathering and processing questions are termed Enquiry Skills questions, unlike other such questions which are based on the evaluation of given resources, these questions basically test your knowledge of a variety of techniques.

◆ In effect you have to learn about these techniques in some detail and be able to discuss the relevance of different techniques to specific situations.

Gathering techniques

The gathering techniques which you must know about are:

1 Extracting information from maps

2 Field sketching

3 Measuring

4 Recording information on a map (land use, location, distribution)

5 Observing and recording (traffic/pedestrian flows, weather, environmental quality)

6 Compiling and using questionnaires and interviews.

◆ There may be other techniques, such as taking photographs, which could be appropriate in some situations. You will still get credit for mentioning them in your answers. At Foundation level you will simply be asked to identify the appropriate technique, usually from a given list.

◆ At General and Credit levels you will be asked to identify appropriate gathering techniques and to give reasons for ('justify') your choice.

◆ At Credit level the techniques are more complex than at General level. You will be expected to give extended answers when identifying the technique; for example, in a river study it is not sufficient to refer to 'measure speed' as it is at General level.

◆ You are expected to give more detail, such as 'measure speed either by using a flow meter or by putting a ball or orange into the river and measuring the time it takes to travel over a measured distance'.

◆ When giving reasons, again you are expected to give more detail in your answer at Credit level. For example, in using a flow meter you might mention the higher degree of accuracy obtained from using this kind of instrument over other techniques. Alternatively, if using an orange, you might refer to the fact that it is a simple and cheap method of obtaining information on river speed.

Processing techniques

The processing techniques which you need to know about are:

1 Classifying, tabulating and matrixing information

2 Drawing graphs (bar, multiple bar, line, pie and scatter)

3 Drawing maps (land use, location and distribution)

4 Drawing cross-sections/transects

5 Annotating or labelling maps, graphs, field sketches, photographs.

◆ As with gathering techniques there may be other techniques which you could mention, such as pictograms, which might apply in certain situations and for which you would be given credit.

◆ When you try to justify the use of these techniques you should discuss how a technique is the right one to use for the data which is being processed. Making comparisons with other techniques in terms of clarity and complexity, for example, using a pie chart as opposed to a bar chart in certain situations, is a good way to justify their use.

◆ Some techniques are more useful than others for showing trends over time (line graphs); for highlighting specific particulars (e.g. labelling field sketches and photographs); and for illustrating relationships (e.g. scatter graphs).

◆ In your answers at Credit level, you are expected to elaborate or give more detail on the techniques, especially when describing gathering techniques. The person marking your paper can award extra marks for detailed description and explanations.

◆ Try to avoid making simple statements such as 'using maps' or 'measuring', even at General level.

◆ You must avoid giving the **same** justification for different techniques because you will not be credited twice for this.

◆ Avoid giving simple statements on processing techniques such as 'easy to read' or 'easy to draw', although you would get credit for a comparative statement such as 'drawing a bar chart is less complicated than drawing a pie chart'. You could go on to explain why this is the case.

Gathering techniques: typical topics

Typical topics examined in questions on gathering techniques include the following:

◆ **Settlement studies:** Changes in land use; Comparisons of different land use zones; Survey of environmental quality; Study of commuter settlements; Sphere of influence studies.

◆ **Industrial studies:** Location of an old manufacturing industry or a modern industrial estate; Changes in industrial areas; The impact of a new industry on an area or decline of an older factory.

◆ **Farm studies:** Changing land use patterns; Land use on a specific farm; Comparison of two or more farms.

◆ **Physical landscape studies:** Comparison of two landscapes; Study of the impact of land use developments such as tourism or industry on a physical environment; River valley(s) study; Footpath erosion.

◆ **Population studies:** Population change in an area; Comparison of population characteristics of two areas/countries.

◆ **Weather studies:** Comparison of the weather at different times or locations; Measuring the weather elements.

Processing techniques: typical topics

Typical data used in these questions include the following:

◆ **Population data:** census details, population growth figures, age/sex data for countries or regions.

◆ **Trade figures:** import and export figures.

◆ **Weather data:** temperatures, rainfall, wind speeds for different locations.

◆ **Farm data:** land use/land use changes or output figures.

◆ **River valleys data:** depth, speed, transects/cross-sections.

◆ **Employment figures:** for different types of industries in different parts of the country.

If you can get access to past papers, look at the different topics examined in both gathering and processing questions to familiarise yourself with the kinds of studies which are included.

Sample Questions, Answers, Comments and Marks Obtained

Questions and Answers

Question: General level 1999 – Gathering techniques

The pupils in Figure I.1 are about to do a river study.

What two techniques could they use to gather information about the characteristics of the river at AB on the sketch?   **(4 marks ES)**

Questions and **Answers** continued   **?**

Figure I.1 Sketch of river landscape

Sample answer

(✓) denotes a correct point.

<u>Technique 1</u>: You could draw a sketch of the river (✓) or take photographs comparing them (✓).

<u>Technique 2</u>: You could then label the sketch and add colour, and compare the photographs.

Justify your choices

By drawing a sketch or taking photographs you could see exactly the characteristics you need to know when drawing the sketch. Labelling and colouring it shows clearly the main characteristics of the river.

Comments and marks obtained

This answer obtains 2 marks for identifying sketches and photographs as field gathering techniques. However, in the remainder of the answer the candidate falls into the trap of confusing gathering and processing techniques. Technique 2 and the justifications really discuss ways of processing the information and give reasons for using these processing techniques.

Had the answer elaborated a little on the comparisons between the sketches and the photographs by referring to the different amount of detail on each

Questions and Answers continued

it may have merited additional marks for justification of the gathering techniques. Total marks: **2 marks out of 4**.

Question: Credit level 2000 – Gathering techniques

A group of students has been asked to gather information about urban and industrial change in the field study area marked on Figure I.2.   **(5 marks ES)**

What three techniques could they use to gather this information?

Explain your choice.

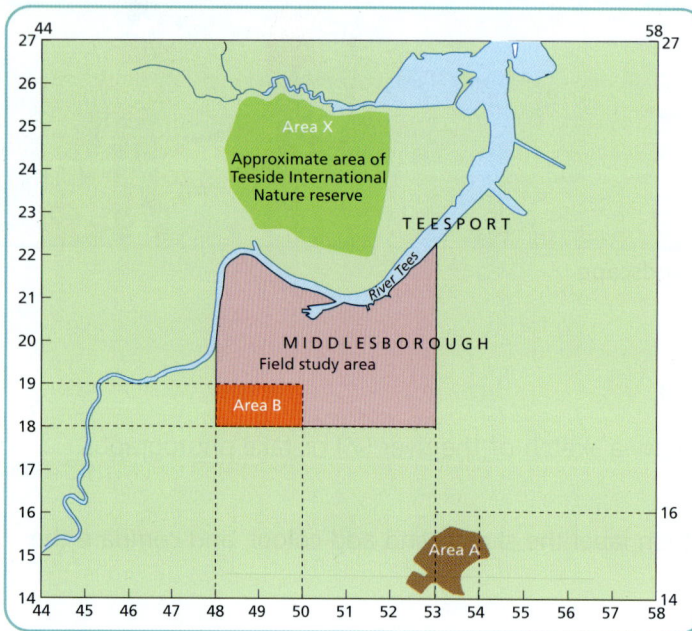

Figure I.2

Sample answer

Photographs taken at different points around this area (✔) will allow the pupils to see the present state of urban and industry (✔) and can be annotated to show how it has changed. An interview with local councillors and businessmen (✔) provides accurate information on urban and industrial change (✔). Books from libraries or even a visit to the local museum (✔) may provide additional information which will be accurate. Drawing a land use map for the present day (✔) would be useful as it can be compared with one from older times found at the library or museum (✔).

Comments and marks obtained

The answer correctly identifies three different gathering techniques: photographs, interviews, and drawing a land use map. Simply stating these without the additional comments would not have gained marks at Credit level. The candidate develops the basic identification very well. Reference to annotation, which is really a 'processing' technique, is not relevant. The justifications are all appropriate, especially the last one which relates the land use map to older data obtained from museums. The mention of books on their own without saying what type of books is not enough for a mark. The answer is good enough for **5 marks out of 5**.

Question: General Level 2000 – Processing techniques

Look at Table I.1.

Give one processing technique which could be used to present the information in Table I.1.

Give reasons for your choice.   **(3 marks ES)**

Table I.1 Japan's exports

Exports	Percentage
Manufactured goods	83
Chemicals and other raw materials	7
Others	10

Sample answer

Technique: Pie chart (✓)

Reasons: This allows a clear comparison to be made and the relative importance is also illustrated. Colour can be used to make the chart more effective.

Comments and marks obtained

The answer gains 1 mark for correctly identifying the pie chart.

The justification part of the answer does not gain any further marks. Although the answer refers to comparisons it does not state what is being compared; nor does it refer to the different segments of the pie chart allowing visual comparison, for example. The next statement correctly discusses colour but again fails to relate this to the different components of the pie chart, thus noting the effective use to highlight differences. In total the answer gains **1 mark out of 3**.

Question: Credit level 1999 – Processing techniques

Look at Tables I.2 and I.3.

Which processing techniques would you use to **compare** the statistics for the two airports?

Justify your choice of techniques.   **(6 marks ES)**

Table I.2 Climatic data for Glasgow Airport, 1–7 September 1997

Date	Average temperature (°C)	Rainfall (mm)	Sunshine (hrs)	Wind direction
1	14·4	0·9	3·7	W
2	12·1	12·6	1·0	S
3	16·8	11·3	1·7	S
4	14·5	10·4	8·3	SW
5	14·7	Trace	4·1	SW
6	13·5	1·6	4·0	SW
7	15·1	0·4	0·0	W

Table I.3 Climatic data for Stornoway Airport, 1–7 September 1997

Date	Average temperature (°C)	Rainfall (mm)	Sunshine (hrs)	Wind direction
1	15·2	0·9	5·0	NW
2	13·2	3·9	0·8	S
3	13·6	8·4	1·1	S
4	12·4	19·1	2·6	S
5	12·0	7·8	0·1	W
6	12·4	5·5	0·4	W
7	13·3	3·8	0·0	SW

Sample answer

Bar charts (✓) and line graphs (✓) could be used to compare the rainfall, temperature and sunshine. A wind rose (✓) could be used for wind direction.

Bar charts are easy to compare and with colour could show similarities clearly (✓). They are used for amounts of something over a period of time such as rainfall and sunshine hours.

Line graphs show changes of something over a period of time (✓). By placing two sets of data on one graph and adding colour it can show comparisons easily.

Wind roses are used to present wind directions in such a way that no real reading of the graph is needed. They are clear and show relevant information and can be annotated to provide additional information.

Comments and marks obtained

Two marks are gained in the first sentence for correctly identifying bar charts and line graphs for rainfall and temperature. A third mark is obtained for stating that the wind rose could be used for wind direction.

A further mark is gained for the statement on adding colour to the bar chart to enhance comparisons.

The statement on the use of bar charts over time is too vague and general to gain any further marks but the reference to line graphs being used to show changes over time is more relevant and gains a further mark.

The second reference to colour being added to the line graphs gains no additional marks since it repeats an earlier statement which was already credited. Also, the references to the wind rose repeat earlier statements and gain no further marks.

The answer gains a total of **5 marks out of 6**.

Maps and Mapping Skills

At all three levels about one-third of the marks available in the paper will be based on an Ordnance Survey map either at a scale of 1 : 50 000 or 1 : 25 000. The questions asked will be a mixture of Knowledge and Understanding and Enquiry Skills and may cover a wide range of Key Ideas, particularly those relating to physical landscape, land use patterns, landscape conflicts, agriculture, industry, settlement and population distribution.

These questions are designed to test your mapping skills. You will have been taught these skills and have been given practice in applying them throughout both years of your Standard Grade course.

There are certain basic skills which you should know and which you should continually practise.

◆ These include being able to give grid references (six-figure references at Credit level), recognise symbols (although the maps will have a key), recognise contour patterns, be able to relate land use to the physical landscape, and understand and use cross-sections.

◆ If you are unable to handle any of these skills, make a point of discussing this problem with your class teacher.

◆ Questions will ask you to give map evidence in support of your answers. Do this by using appropriate grid references to locate relevant features and refer to features on the map to show that you are actually using the map in your answer. It is advisable to give a number of grid references in your answer to be safe.

◆ Some questions will ask for descriptions of landforms, features and patterns, and identification of certain physical and human features. Others will ask for your opinion, based on map evidence, on how various features interact with each other. For example, you may be asked about the advantages and disadvantages of a particular site perhaps for a farm, settlement or industry. Here you are combining your knowledge of these topics with mapping skills.

◆ In many questions you may be asked to **describe** or **explain**. As noted earlier, it is vitally important that you do not confuse these instructions. If asked to *explain* you must use phrases such as 'due to', 'this is because' and 'as a result of'. If you do not, you will lose important marks in your answer, even if you provide a long, well-written answer.

◆ The length of your answer and the detail given should be related to the number of marks available. More detail is usually required for full marks in a Credit level answer than in a General level answer.

◆ Do not worry if the map is different from those you have used in class. Each map will have a key which can help you to answer map-based questions. However, it helps enormously and saves you precious time if you are already familiar with most map symbols.

◆ All of the answer is there on the map. If you read the questions carefully, refer to the map and make use of the map key, you should score high marks.

◆ Again, if you get to the end of the examination and find that you have some time left, go back to the map question and go over your answers, adding additional detail and evidence to any answers which may be a little short.

◆ You may be asked to provide labelled diagrams, for example, if you are asked to explain the formation of certain physical landforms which you have identified on the map.

◆ Clearly labelled, well drawn diagrams can often obtain full marks. Practise this skill when revising map-based questions.

Sample Question, Answer, Comments and Marks Obtained

Questions and Answers

Question: General level 2003

Describe the physical features of the River Clyde and its valley between 774530 and 737560. **(4 marks KU)**

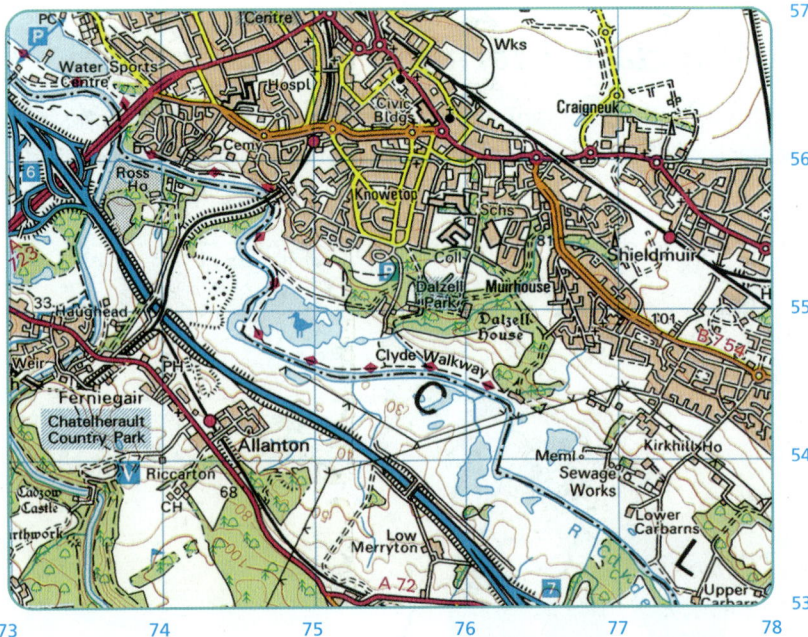

Figure I.3

Sample answer

At 745555 there is clearly meandering present (✓). This means that it is in the middle or lower course (✓). The river starts with quite a big width and decreases slightly at the end (✓). Tributaries join at 761544 (✓) and there are ox-bow lakes (✓) at 758542 (✓). The river's direction is southeasterly (✓).

Questions and *Answers* continued ➢

Comments and marks obtained

This is an excellent answer which answers both parts of the question – that is, it describes both the physical features of the river and its valley. Often candidates answer only one of these, either the river or its valley. The answer correctly refers to meanders, tributaries, the stage of development, ox-bow lakes and direction of the river's flow, and gives good grid references to support the answer. The question is worth **4 marks out of 4**, but this answer could gain much more than this as it contains additional correct detail.

Glossary of Associated Terms

The words contained within the Glossary sections are **key words** which you should know in relation to each of the Key Ideas and be able to use in your answers when appropriate.

There is a Glossary at the end of each of the Key Ideas sections.

THE PHYSICAL ENVIRONMENT

Key Idea 1

Physical landscapes are the product of natural processes and are always changing.

Key Point 1

At all three levels – Foundation, General and Credit – the landscapes which you have to study include those of glacial erosion and river valleys. You have to be able to recognise the main features of these landscapes, be able to describe them and explain how they were formed. These landscapes are studied within the context of the United Kingdom.

It is important that you fully understand the term 'natural processes'.

Natural processes include the effects of **physical and chemical weathering** on the landscape and the impact of **erosional** and **depositional** processes which contribute to change in the landscape. Physical weathering refers to the impact of elements of the weather such as temperature changes, rainfall, wind, water flowing across the land in the form of rivers and streams, and a process which happened many thousand of years ago – **glaciation**.

About one million years ago many parts of Britain were covered by large sheets of ice called glaciers. As these ice sheets moved, they altered the landscape over which they passed. Two things happened. First, the ice sheets eroded – that is, they wore away the landscape, carving out new features – and secondly the glaciers transported eroded material and deposited it in other parts of the country.

You should be able to identify the main features created by both erosion and deposition in those parts of Britain which have been subjected to glaciation.

At all levels you should be able to identify these features on a diagram, such as that shown in Figure 1.1.

On a landscape diagram or on an Ordnance Survey map, either on the scale 1 : 50 000 or 1 : 25 000, the features which you should be able to recognise from the contour patterns include features formed by erosion such as: pyramidal peaks, corries, corrie lochs/lakes/tarns, hanging valleys, U-shaped valleys, truncated spurs and arêtes.

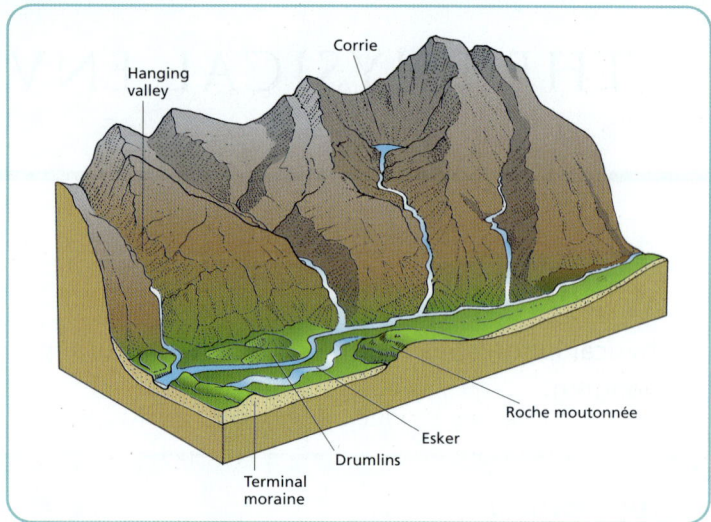

Figure 1.1 Features of upland and lowland glacial landscapes

Features formed by ice sheets depositing material include various types of moraine, outwash plains, eskers, drumlins and erratics.

Key Point 2

You are expected to know the main features of these landscapes and how they were formed.

Glaciated Uplands

During the last 2.5 million years the British Isles have been covered at different times with large sheets of ice. These periods are called glaciations and there may have been up to 20 different glaciations during the period known as the Ice Age. The ice advanced southwards as the climate became colder and retreated northwards as the temperature became warmer. These advances and retreats are termed glacial and interglacial periods respectively.

Formation processes

1   Glacial erosion occurs through two processes called abrasion (a sandpapering effect as the ice moves across the land) and plucking (where pieces of rock are torn away from the land). Abrasion produces smoothed surfaces and plucking tends to produce jagged features.

2   The glaciers themselves consist of ice in which melting ice near the base of the glacier causes a process within the glacier known as internal deformation.

3   The weight of the ice causes the glacier to slide on top of the melting ice. This is called basal sliding. The rate of flow of the glacier depends on the type of rock over which it flows, the amount of ice in the glacier and the slope of the land. This process of melting is called ablation.

4   As glaciers move they erode and deposit material at their margins – that is, at the front and sides. Meltwaters flowing from the glaciers further erode and deposit material in a process known as the fluvioglacial process.

Key Point 3

Erosion by glaciers produces several landforms. You should be able to identify and describe their formation and their characteristics.

Landforms

U-shaped valleys

As a glacier moves downhill through a valley, the shape of the valley is transformed. A material called boulder clay is deposited on the floor of the valley. As the ice melts and retreats the valley is left with very steep sides and a wide, flat floor. A river or stream may flow through the valley due to meltwater from the glacier. This replaces the original stream or river and is known as a misfit stream. If the material which is pushed in front of the glacier is left, this material is called terminal moraine. The valley dammed by the moraine may then flood creating a lake which may twist and turn, so it is termed a ribbon lake.

Hanging valleys

The sides of the U-shaped valley are usually high and steep. During the Ice Age, tributary valleys often had smaller glaciers. The glacier in the main valley cut off the bottom slope of the tributary valley leaving it high above the main valley. Today, tributaries of the main valley plunge from the slopes of the main valley straight into the bottom of the valley. These smaller valleys are called hanging valleys.

Truncated spurs

A spur is the bottom part of a slope which juts out into the main valley. As the ice cuts through the original valley, the original spurs are removed by the ice. The feature that remains once the ice melts is called a truncated spur.

Corries, arêtes and pyramidal peaks

Corries, also known as cwms and cirques, are steep-sided hollows in the sides of mountains

where snow has accumulated and gradually turned to ice. The movement of ice in the hollow causes considerable erosion both on the floor and on the sides of the depression. The erosion on the floor is caused by abrasion and the floor becomes concave in shape while the edge takes on a ridge-shaped appearance. Plucking of rocks takes place at the sides as the ice moves forward and the back wall of the depression becomes very steep. Eventually, as the corrie fills up with ice it cannot contain any more and some of the ice moves down the slope to a lower level. Sometimes as the ice melts the meltwater fills the corrie forming a corrie lake – also called a corrie loch in Scotland and a tarn (in England).

Occasionally corries develop on adjacent sides of a mountain and when they are fully formed they will be separated by a knife-shaped ridge termed an arête. If corries develop on all sides of a mountain, the arêtes will form a jagged peak at the top. This feature is called a pyramidal peak. Arêtes and pyramidal peaks are further sharpened by frost action.

Key Point 4

You should be able to provide a sketch to show features of glacial erosion and be able to explain how the landscape with these features was formed.

Study the sketch in Figure 1.1 and learn to recognise the various features shown. If asked how the landscape was formed you should refer to the processes of glacial erosion. At General and Credit levels you may be asked to provide an explanation of how specific features were formed and perhaps provide appropriate diagrams showing the formation process.

Features of glacial deposition

Moraines
Moraine consists of material which has been eroded and transported and deposited by the glacier. This material may be dumped at the end or snout of the glacier and is then called terminal moraine.

Erratics
These are large boulders which have been lifted, carried and deposited by the glaciers in a different part of the country. The rock type of the erratic is usually different from that of the area in which it is deposited.

Outwash plains

These are gently sloping plains consisting of sands and gravel. They are deposited by meltwater streams flowing out from the ice sheet and carrying material collected by the glacier.

Eskers, drumlins and kames

◆ **Eskers** are elongated ridges of coarse, stratified, fluvioglacial sands and gravels and are thought to have been formed by meltwater flowing through tunnels within the lower parts of the glacier and depositing the material.

◆ **Drumlins** are oval-shaped mounds which can be up to 100 metres high and have a 'basket of eggs' look to them. The material in them was deposited due to friction between the ice and the underlying rock causing the glacier to drop its load.

◆ **Kames** are irregular-shaped mounds of material consisting of sands and gravel again laid by glacial streams. Sometimes they are formed as terraces between the glacier and the side of the valley wall, forming low hills.

Key Point 5

You should be able to describe the relationship between a glaciated landscape and land use within that area.

Glaciation features have an impact on how the land is used for activities such as farming, forestry, communications, settlement, industry, water storage, tourism and recreational purposes. You could describe the limitations on, for example, the use of land for arable farming, the suitability of steep slopes for forestry, or the use of U-shaped valleys for communications and settlement.

Types of Questions

◆ You can be asked to describe how these features were formed but the level of detail which you need to put into your answer obviously depends on the level of the examination which you are attempting.

◆ You will very rarely be asked to explain the formation of these landscape features at Foundation level. However, this type of question is often set in both the General and Credit papers.

◆ If the Ordnance Survey map is based on one of the upland glaciated areas of Britain, e.g. Snowdonia, the Cairngorms or the Lake District, the question may offer a choice of features and ask you to select one and explain in detail how it was formed.

Types of Questions continued ➤

Types of Questions *continued*

◆ This kind of question is also a very popular choice for Question 2 at both General and Credit levels and is often based on a block diagram such as that shown in Figure 1.1.

◆ You may also be asked to provide appropriate diagrams in your answer. It is quite possible to gain full marks for a well-annotated (labelled) series of diagrams.

◆ Processes involved in the formation of these landscape features which you could refer to include abrasion, plucking, rotational movement of ice, transportation and deposition.

◆ Your Standard Grade class notes and textbooks should have already provided you with detailed description and explanations of how the features mentioned above were formed.

◆ The following questions and sample answers should give you an insight into the kind of responses which would merit full marks in a General paper and a Credit paper on this topic.

Sample Questions *and* Answers

Question: General level 2000

Explain how one of the glacial features (arête, hanging valley, U-shaped valley) is formed.   **(3 marks KU)**

Sample answer

An arête is formed when two glaciers are about to meet (✓). No abrasion is used to form this, only plucking and frost shattering (✓). The arête is usually very steep with scree at the bottom.

Comments and marks obtained

Without the diagrams showing the processes and stages of development, worth a mark in itself, this answer would only have gained 2 marks, as indicated. The last sentence *describes* rather than *explains* and is therefore not worth any further marks. With the diagrams the answer is worth **3 marks out of 3**.

Sample Questions and Answers continued ➤

Sample Questions and Answers continued

Question: Credit level 2001

pyramidal peak, hanging valley, arête, corrie

Explain how one of the features listed above was formed. You may use diagrams to illustrate your answer.   **(4 marks KU)**

Sample answer

Ice and snow gather in a hollow (✔) and as more gathers the bottom layers are compressed by the weight (✔) and begin to erode the hollow (✔). When the ice and snow finally melt, a steep back wall is left and the original hollow is much deeper (✔). As there is nowhere for the water to drain away to, a loch called a tarn is formed in the corrie (✔).

Comments and marks obtained

This answer is quite basic but it contains enough detail, especially about the ice and snow being compressed and eroding a steep back wall and melting ice left to form a corrie-loch or tarn. A better answer might have referred to 'plucking and abrasion' and the rotational movement of the ice in the hollow helping to create the back wall and a 'lip' at the front edge of the corrie. However, the answer manages to make enough points to obtain **4 marks out of 4**.

Rivers and their Valleys

Key Point 6

As with glaciated landscapes you should be able to identify the main features of a river and its valley from either an Ordnance Survey map or an appropriate diagram.

In addition to identifying these features at all three levels, at General and Credit levels you could also be asked to both describe and explain in detail their formation.

Features and stages of rivers and their valleys

You should know that river valleys can exist in any one of three main stages of development: upper (youthful), middle (mature) and lower (old age).

◆ Features associated with the **upper** course/stage include:

V-shaped valleys, fast-flowing streams, narrow valleys, steep gradients, waterfalls.

The process taking place is mainly that of erosion.

◆ Features associated with the **middle** course/stage include:

gentler gradients, the river flowing more slowly than in the upper stage, a wider valley with less steep sides, and the river beginning to wind or meander at various points. At this stage the river both erodes its valley and deposits material further downstream in its meanders or when it floods both sides of the valley. Thus it creates a feature called its floodplain.

◆ Features associated with the **lower** course/stage include:

the river valley becoming much wider, the gradient much gentler, the river speed much slower and deeper, and features such as meanders, ox-bow lakes, braiding, wide, flat floodplains, confluence points and river estuaries.

These features are shown in Figure 1.2.

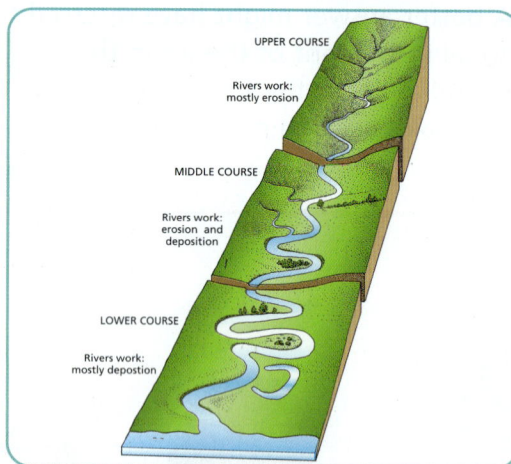

Figure 1.2 Features of a river valley

Types of Questions

◆ Many of the questions set on rivers, especially at General and Credit levels, are based on Ordnance Survey maps. Typically a question may be part of Question 1 which is usually based on the Ordnance Survey map.

Types of Questions continued ➤

Types of Questions *continued*

◆ The question may ask you to locate a section of a river by giving two six-figure grid references.

◆ The question may then ask you to describe the main physical features of this section of the river and its valley.

◆ A common error is that candidates answering this type of question refer to human features such as bridges, roads, houses or industries. These are irrelevant since the question has specifically asked for physical features.

What Should You Describe?

◆ In terms of the river you should refer to the direction of flow, its width and speed (which is usually worked out from the gradient).

◆ In describing the valley features you should mention details such as the stage of development, the width of the valley, whether there are features such as steep slopes, floodplains, meanders, braiding, ox-bow lakes.

◆ Occasionally the question may ask you for an opinion on land use within the valley area, such as whether the area would be a suitable site for a settlement or whether the valley creates any difficulties for communications and, if so, how these problems might be overcome.

◆ These are Enquiry Skills questions.

◆ Another frequently asked question which tests Enquiry Skills would be to ask about the kinds of gathering techniques which could be used in a field study based on a section of the river.

◆ The techniques which you could refer to might include:

drawing field sketches, taking photographs, or measuring different aspects of the river and the valley, such as its width, speed and depth.

◆ Note carefully that in referring to these techniques at Credit level you should include much more detail of how the techniques are applied than in an answer to a General level question. For example, you would describe how the depth of the river was measured using a meter stick. At Credit level the answer would be developed to explain that depths would be taken at regular intervals across the river.

◆ You will also be required to justify your use of the techniques you have mentioned. *Justify* simply means 'give reasons for'.

What Should You Describe? continued ➢

What Should You Describe? *continued*

◆ These techniques might include using a flow meter in order to gain accurate readings. That is certainly more accurate than using an orange or ball over a measured distance.

◆ Using the orange/ball method is useful when there is no access to instruments, which can be expensive.

◆ Taking photographs can show exactly what the valley looks like at a certain point in time. Taking photographs either at different times or in different parts of the valley would allow comparisons to be made at different stages of the river or under different conditions.

Sample Questions *and* Answers

Question: General level 2003

Describe the physical features of the River Clyde and its valley between 774530 and 737560. **(4 marks KU)**

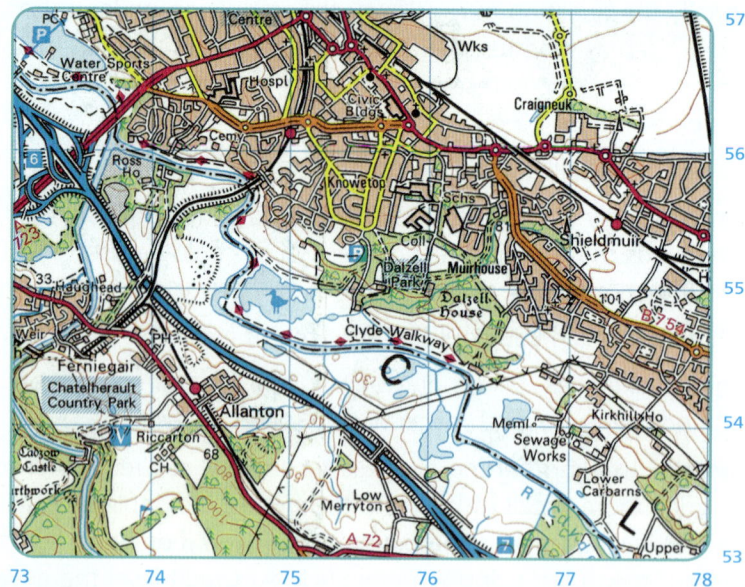

Sample answer

The river is located in a U-shaped valley. The river is heading NW (✔) and changes direction (✔) to N and back again (762544) (✔). The river meanders (✔) at (746555).

Sample Questions and *Answers continued* ➤

Sample Questions and **Answers** continued

Comments and marks obtained

The first statement is inaccurate and gains no marks. The answer correctly notes the direction of flow, a change in direction and a river feature, namely the meander, and gives two correct grid references, gaining a mark for one of these grid references. Notice that this answer refers only to physical features of the river and its valley and does not refer to human features such as bridges, roads or settlement which would have been inaccurate.

The answer therefore gains the full **4 marks out of 4**.

Question: Credit level 1999

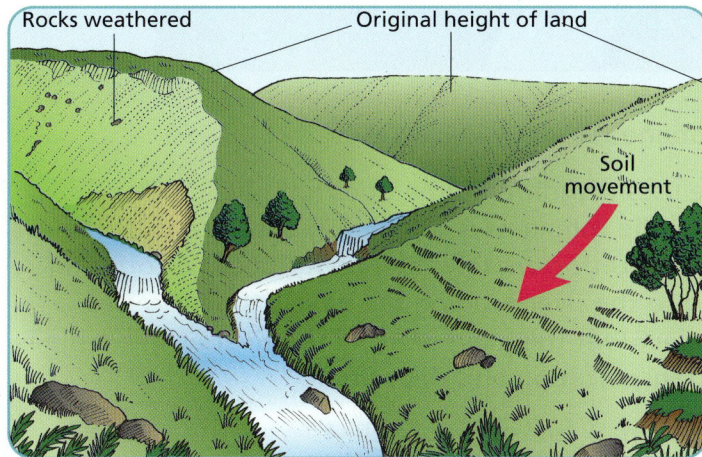

Figure 1.3 Field sketch of the upper course of a valley

Look at Figure 1.3 above.

Explain how the V-shaped valley was formed. **(4 marks KU)**

Sample Questions and **Answers** continued ➤

Sample Questions and Answers continued

Sample answer

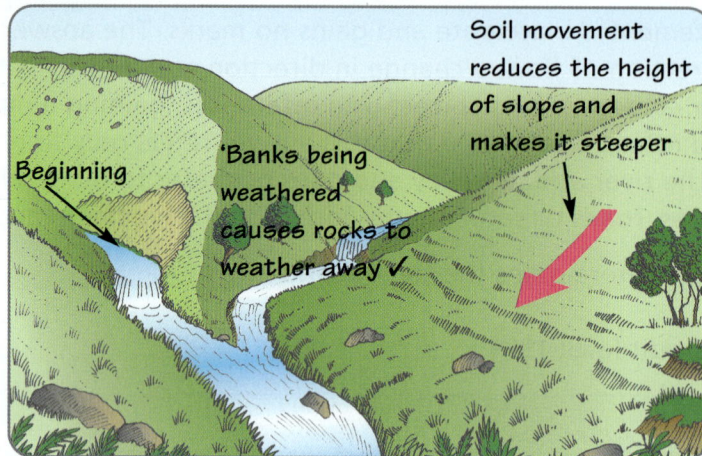

Beginning

'Banks being weathered' causes rocks to weather away ✓

Soil movement reduces the height of slope and makes it steeper

Comments and marks obtained

Although the candidate uses diagrams he/she does not give much information. Only the **middle comment** is sufficiently accurate to obtain a mark for 'weathered away'. Unfortunately the answer does not offer an explanation by referring to the river eroding the valley, the processes involved or even to the fact that the main work of the river at this stage is erosion as opposed to transportation and deposition.

This answer obtains only **1 mark out of a possible 4**.

Glossary of Associated Terms
Key Idea 1: Physical Landscapes

Abrasion: the process by which rocks within ice sheets and rivers scrape and erode the land over which they pass.

Alluvium: the material deposited by a river, usually over its floodplain.

Arête: a narrow ridge between two corries created as the corries are formed on two adjacent sides of a mountain.

Attrition: process of rocks in a river wearing away by constantly rubbing together.

Braiding: process by which rivers divide into separate channels through material being deposited by the river in midstream.

Corrie: an armchair-shaped hollow on the side of a mountain formed by ice filling a hollow and eroding the side of the mountain by abrasion and plucking and rotational movement at the base of the hollow. When the glacier melts a lake or loch is sometimes left, called a corrie loch, lake or tarn.

Drift: material deposited by a glacier. It is made up of two main parts: 'till' which is deposited under the glacier, and 'outwash' which is formed by meltwater streams carrying away particles of material from the debris under the glacier.

Distributary: a branch of a river flowing out from the main river.

Drumlin: an oval-shaped hill formed from deposits within a glacier.

Erosion: process by which rocks and landscapes are worn away by agents such as moving ice, wind, flowing water and sea/wave action.

Erratic: rock or boulder which has been moved by ice during the Ice Age from its original location and left in another part of the country.

Esker: long ridge of sand and gravel deposited by rivers which flowed under ice sheets.

Estuary: the mouth of a river where the river meets the sea.

Floodplain: the area on either side of a river, usually in its middle or lower stage, which is formed from material carried by the river and deposited when a river overflows during a period of flooding. Floodplains can vary greatly in size depending on the stage of the river. The deposited material is called alluvium.

Freeze–thaw action: this happens when water trapped in cracks in rocks alternately freezes and thaws causing the rock to break up.

Frost shattering: similar to freeze–thaw. It is caused by water turning to ice and as it melts it expands, putting pressure on the rock and eventually causing it to shatter. Material from this may fall to the bottom of slopes and gather as rock debris known as scree.

Glacier: a large mass of moving ice which changes the shape of the land over which it passes.

Hanging valley: a valley which is situated at the side of a U-shaped valley. It is a tributary of the main valley.

Mass movement: process by which rocks move under gravity.

Meander: bend formed in the middle or lower course of a river.

Moraine: material which is deposited by glaciers. Different types include 'end moraines' or 'terminal moraines' which are formed at the front of the glacier as it melts, 'lateral moraines' formed at the sides of glaciers, and 'medial moraines' formed in the middle of glaciers or at the edges of where two glaciers meet.

Ox-bow lake: lake formed from a former meander when the loop of the meander is cut off from the main channel by a build-up of river deposits.

Plucking: process by which moving ice tears rocks from the surface over which it moves.

River cliff: steep bank of a river formed as the river undercuts or erodes the outside bank.

Source: the point at which a river begins.

Till: material deposited beneath and at the end of a glacier. It consists mainly of boulder clay.

Transportation: process by which rock particles are carried by rivers or glaciers or wind.

Tributary: a small river which joins with a larger river.

Truncated spur: a piece of land the bottom of which at one time jutted into a valley and was cut away or eroded by a glacier flowing through the valley.

U-shaped valley: a valley with very high, steep sides and a wide, flat bottom formed by a glacier flowing through the original valley. It is now usually occupied by a small river which is known as a 'misfit stream' because it is not the original river that once flowed through the valley.

V-shaped valley: a narrow valley which is typical of a valley in the upper or youthful stage of its development.

Waterfall: a steep break in the course of a valley often associated with a change in rock type along the course of a river.

Weathering: process by which rocks are worn away through physical action such as flowing water, wind, or a chemical reaction between rocks and rainfall which may have become acidic.

Key Idea 2

The elements of weather can be identified, observed, measured, recorded and classified. As a result, dynamic patterns can be identified and used for forecasting.

Key Point 7

You should know the various weather elements, the methods and instruments used to measure them and all of the units in which they are recorded.

Weather elements are the various features which contribute to weather. Weather is the day-to-day conditions of the atmosphere.

The weather elements, weather instruments and units of measurement are shown in Table 1.1.

Table 1.1 Weather elements, methods of measuring and units of measurement

Element	Method of measuring	Units
Precipitation (all forms of moisture in the atmosphere)	Rain gauge	Millimetres
Temperature	Maximum and minimum thermometers	Degrees centigrade/Celsius (°C)
Humidity	Wet and dry bulb thermometers	Relative percentage
Sunshine	Sunshine recorder	Hours of sunshine
Cloud cover	Observation	Eighths of the sky – oktas
Wind speed	Anemometer	Knot miles/kilometres per hour
Wind direction	Wind vane	Points of the compass
Visibility	Observation	Kilometres or parts thereof
Air pressure	Barometer	Millibars

Weather Instruments

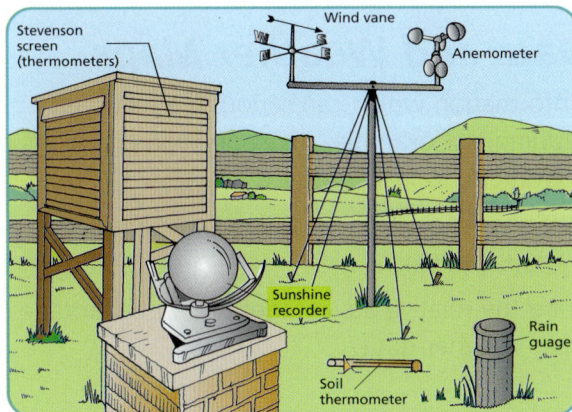

Figure 1.4 Weather instruments

◆ The thermometers are kept inside a white box with slatted sides called a Stevenson screen. You should know the best place to position this box: out in the open away from buildings and trees.

◆ You should also be aware of how these instruments work.

◆ There are other methods of recording weather data besides those listed above. These include using radiosonde balloons, aircraft, ships and satellites.

◆ Information from these sources is gathered every day and sent to meteorological offices where the data is analysed and recorded on weather maps.

Weather Maps

◆ These maps are constructed with lines, called isobars, which join up places with the same pressure. The pattern of the isobars indicates the type of pressure system which is present.

◆ Those patterns which have isobars close together are low pressure areas or depressions.

◆ Where the isobars are widely spaced apart these show high pressure systems or anticyclones.

◆ You need to understand these patterns and know the kind of weather associated with them in order to answer questions that ask you to explain the weather and weather changes.

◆ This information is passed on to newspapers, television and radio companies for publication.

The Use of Weather Data and Weather Maps

◆ Various people require weather information which can seriously affect their work or day-to-day decisions including, for example, transport workers, pilots of aircraft, ships' captains, building workers, tourists, farmers, etc.

◆ You may occasionally be asked at Foundation or General level to explain the importance of weather data to various people.

◆ You could also be asked to describe the relative advantages and disadvantages of the different means of gathering weather data.

Recording Weather Data

◆ Weather data can be recorded in different ways, including weather maps with various symbols, such as those published in newspapers and shown on television.

◆ Information for individual weather stations can be shown using a variety of symbols drawn around a weather circle.

◆ At both General and Credit levels you should know these symbols and be able to describe weather conditions from given weather stations such as that shown in Figure 1.5.

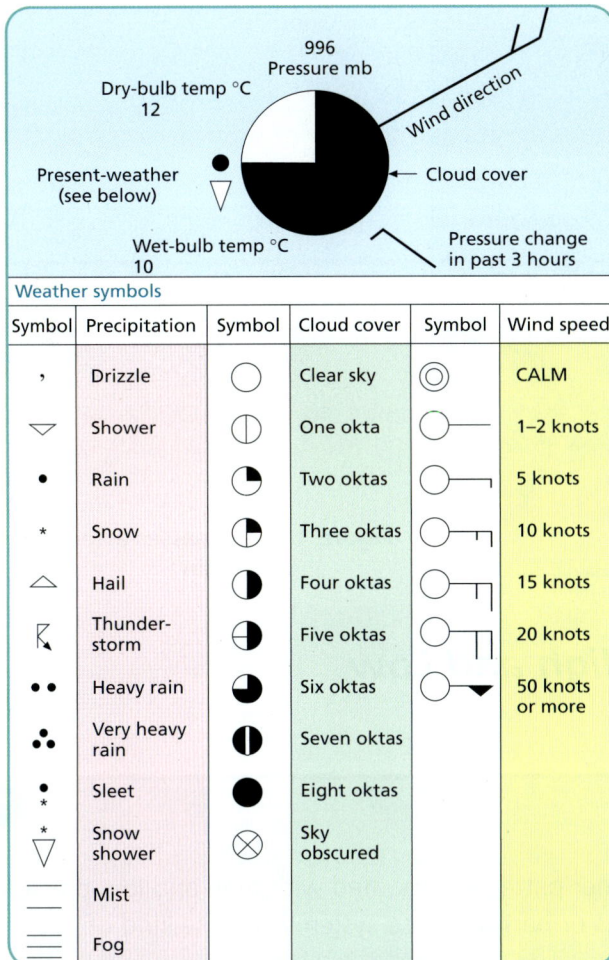

Figure 1.5 A weather station

Weather Forecasting

Weather forecasting is based on knowledge of patterns of air pressure as recorded on weather maps, which are sometimes referred to as synoptic charts.

You should know about weather systems such as **low pressure areas (depressions)** and **high pressure systems (anticyclones)**.

◆ Britain's weather is greatly affected by these systems throughout the year.

◆ These systems result from the movement of large bodies of air called air masses which originate in different parts of the world.

◆ These masses may originate in warm, cold, wet or dry regions and travel towards the British Isles.

◆ Often cold and warm air masses collide to create areas of low pressure which generally flow across Britain from west to east.

◆ Figure 1.6 shows the main air masses which affect Britain's weather.

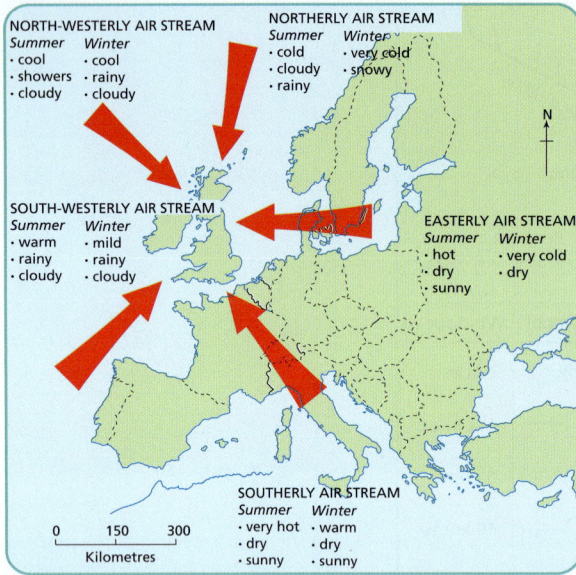

Figure 1.6 Air masses affecting Britain

Weather Systems: High and Low Pressure Systems

Key Point 8

You must know the various weather patterns associated with high and low pressure systems, the different components or parts of these systems and the kind of weather they bring.

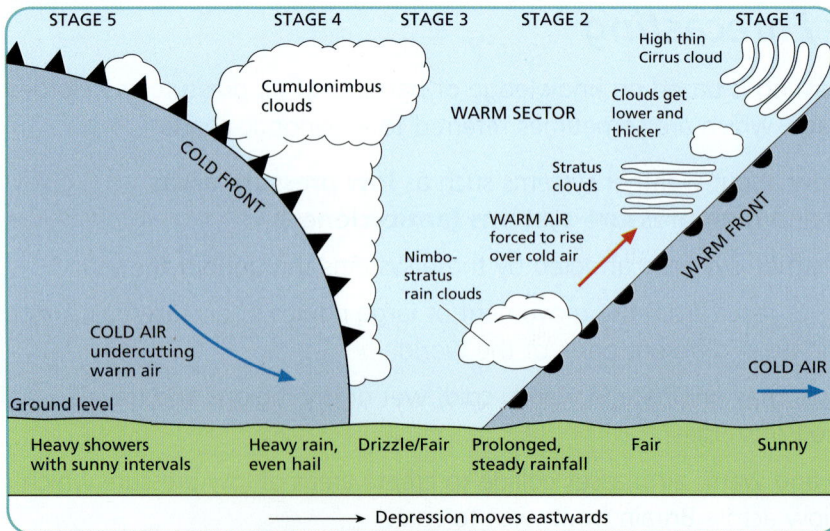

Figure 1.7 The passage of a depression

◆ Generally low pressure systems bring wet and windy weather. Depending on the time of year and where they come from, depressions can, for example, lower temperatures in summer, bring a thaw during a severe cold spell in winter, or bring snow and sleet from northern latitudes.

◆ In order to explain weather changes caused by a depression, you should understand the different parts of low pressure systems and the type of weather linked to them (Figure 1.7).

◆ These features include:

– **warm fronts** – increasing winds, cloud cover, drizzly rain and slight rise in temperature

– **cold fronts**, which lag behind the warm front bringing much heavier winds, large clouds, heavy rain and lower temperatures

– **occluded fronts**, which are found near the centre of the low pressure area with mixed conditions of rain, wind and cloud cover

– the **warm sector**, which is found between the warm and cold fronts – conditions tend to settle, skies are dull and cloudy, temperatures rise slightly, there is little rain.

◆ If you are asked to **explain** weather conditions you must use phrases such as 'this happens because' or 'due to' or 'as a result of' and then refer to the appropriate weather system and the fronts which are passing.

◆ Notice that in examination questions the word **explain** will always be in bold type to draw your attention to what you have to do in your answer to gain full marks.

◆ If the question is based on Enquiry Skills you may be asked to express an opinion on weather patterns and their effects based on a given resource, e.g. a weather map or maps.

◆ You need to use the data provided in the resource to support your answer. You will gain 1 mark for each correct statement.

◆ Anticyclones are areas of high pressure which bring calm, bright, sunny conditions. However, the temperatures depend on the time of year:

– in winter, anticyclones can bring very low temperatures which can be below freezing during the night, causing extreme frosty and occasionally foggy conditions

– in summer, high pressure systems can bring periods of very warm and sunny weather

◆ Your course notes, workbooks and textbooks will contain great detail on weather elements, instruments, weather systems and their component parts.

Sample Questions and Answers

Question: General level 2000

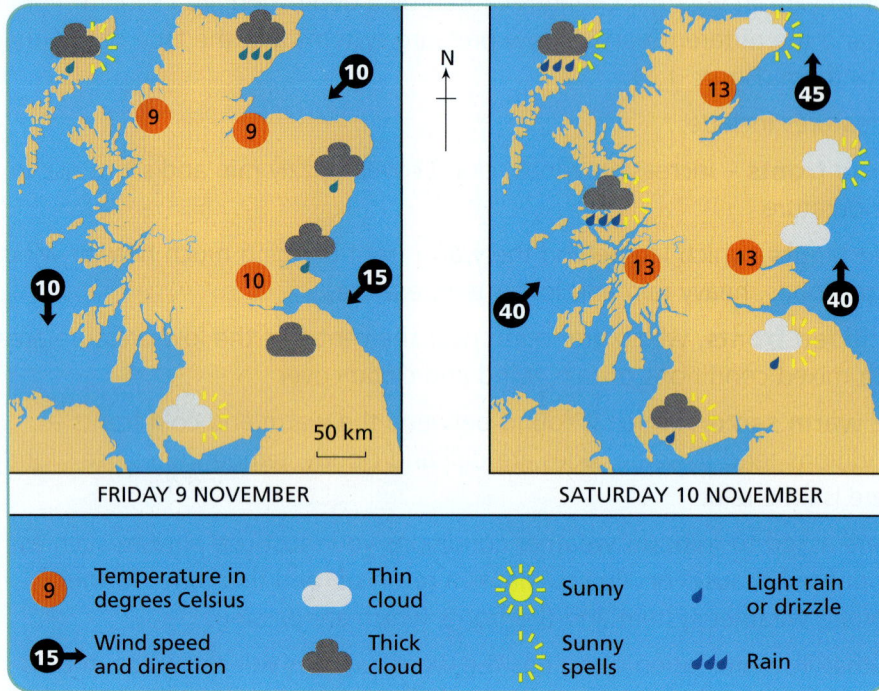

Figure 1.8 Weather conditions in Scotland, November 1995

Look at Figure 1.8.

Compare Friday's weather with Saturday's weather along the east coast of Scotland. You must refer to more than one weather element. **(4 marks ES)**

Sample answer

The wind speed is a lot faster on Saturday (✔) and it is coming from the north. There are sunny spells with light drizzle but on Friday there was thick cloud (✔) with light drizzle (✔). There were higher temperatures on the Saturday (✔).

Comments and marks obtained

The answer correctly mentions more than one element. It also follows the instruction to compare Friday and Saturday's weather instead of simply describing the weather on both days. Comparisons of wind speed, cloud cover, rainfall and temperature gained the full **4 marks out of 4**.

Sample Questions and **Answers** continued ➤

Sample Questions and **Answers** continued

Question: Credit level 2000

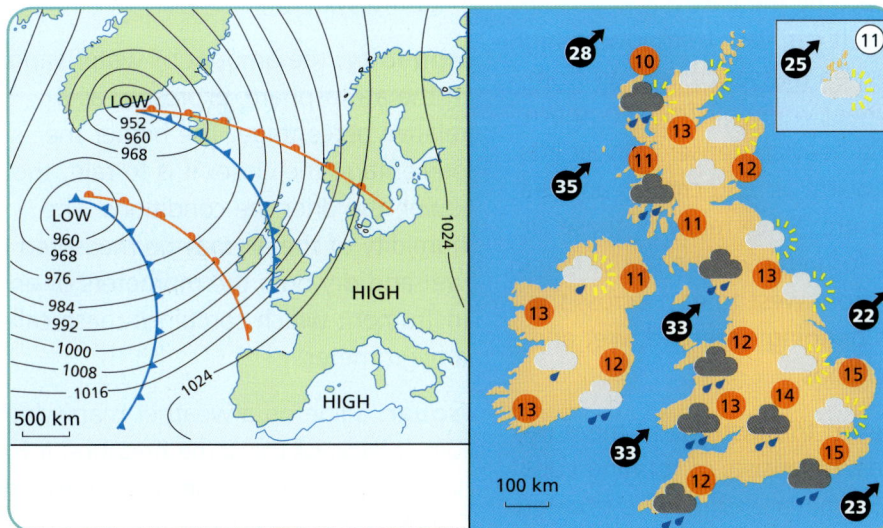

Figure 1.9 Weather conditions over the British Isles, October 1998

Look at Figure 1.9.

Explain the differences in weather between the east and west of the British Isles. **(6 marks KU)**

Sample answer

The east of Britain is under the warm sector of the depression (✓). This brings some cloud and sunny spells (✓) and temperatures which are slightly higher than the surrounding air (12–15 degrees as shown on the weather map (✓)). The west, however, is under the cold front which brings heavy cloud (✓) and heavy rain. The temperatures also start to drop (10–13 degrees on the map). In Ireland, which is in the cold air behind the cold fronts, clouds are thinner with some rain (✓) left by the cold front as it passes and temperatures are still slightly lower (✓) than in the warm sector (11–13 degrees).

Comments and marks obtained

This is an excellent answer which refers to all of the main elements of the weather which are relevant. Good comparisons are made throughout and good use is made of the data. The answer also includes good explanatory points and, as the ticks indicate, there are more than sufficient correct statements for this answer to gain the full **6 marks out of 6**.

Glossary of Associated Terms
Key Idea 2: Weather

Air pressure: the pressure put on the Earth's surface by the Earth's atmosphere. It can vary from low to high pressure. When masses of cold air from the north meet warm air from southern latitudes the heavier, denser cold air lifts the lighter warm air and creates an area of low pressure. Since many of these air masses are formed over the sea or ocean they move from west to east across the British Isles bringing unsettled, wet and windy conditions.

Anemometer: this instrument measures wind speeds. It consists of rotating cups and a speedometer and may be situated on top of a high pole. A smaller hand-held version can also be used.

Anticyclone: an area of high pressure which can remain in place for fairly long periods of time.

Barometer: this instrument measures air pressure in millibars. A drum covered by graph paper is attached to a pen which records or traces changes in air pressure over a specific period, e.g. a week.

Beaufort scale: a scale for measuring wind speed by observation, e.g. calm, breezy, stormy, and ranging from 0 to 12.

Celsius scale: an alternative name for measuring temperature in degrees (°) centigrade.

Cold front: the boundary between cold air and warmer air within a depression. It is usually associated with stormy and very wet conditions.

Depression: another name for a low pressure system.

Humidity: the amount of water vapour in the atmosphere, expressed as a relative percentage. The higher the percentage the closer it is to rain; the lower, the drier the conditions. The humidity of the air may be measured by wet and dry bulb thermometers or an instrument which combines these called a hygrometer.

Isobar: a line on a weather map which joins places of the same pressure. It is similar to a contour line on an Ordnance Survey map.

Maximum thermometer: this instrument allows readings to be taken of the maximum temperature reached in any given day.

Minimum thermometer: this instrument is filled with alcohol rather than mercury. It allows the reading of the lowest temperature reached in any given day. By adding the maximum and minimum temperatures together and dividing by two we get the average daily temperature.

Occluded front: this occurs where a cold front overtakes a warm front, usually near the centre of a depression.

Okta: when describing the amount of cloud cover the sky is divided into eighths. One-eighth is called an okta.

Precipitation: this refers to all forms of moisture in the atmosphere, e.g. rain, snow, sleet and hail.

Rain gauge: the instrument used to collect rainfall during a 24-hour period.

Stevenson screen: a wooden box, painted white with slatted sides, in which weather instruments are placed.

Sunshine recorder: an instrument for measuring the number of hours of sunshine in a day. A piece of paper is placed behind a glass ball and the sun shining through the ball burns a mark on the paper.

Synoptic chart: a map of an area showing isobars, fronts and pressure systems. Weather forecasts can be made from study of this map.

Temperature: the amount of heat in the atmosphere.

Warm front: the boundary between the edge of the warm air in a depression and the air which it is replacing. This is usually associated with persistent drizzle and increasing wind and cloud cover.

Warm sector: an area of warm air between the warm and cold fronts in a depression.

Weather station: a site where instruments are set up to measure the various elements of the weather.

Weather station circle: on a weather map, a circle surrounded by various symbols. These represent the weather conditions at a particular weather station.

Key Idea 3

The world can be divided into major climatic zones.

Key Point 9

The climate of any area is based on the average weather conditions recorded over a period of time which can vary from a year to several years. Two features of weather are recorded and shown on a graph: total monthly rainfall from January to December, and average monthly temperatures throughout the year.

Key Point 10

For the purpose of the examination you need to know the climates of four main zones, namely: Tundra, Mediterranean, Hot Desert and Equatorial.

Climate Graphs

◆ In the examination you will be asked to identify the four climate types listed above. At General and Credit levels you may also be asked to describe the climates in detail and perhaps comment on the impact of the climate on local people and environments.

◆ When describing climate graphs, refer to rainfall distribution throughout the year, for example, the wettest and driest times. Also quote amounts, such as the highest and lowest, and describe whether the climate is relatively wet or dry, noting exceptional conditions.

◆ Having described the rainfall pattern, describe the temperature pattern noting the highest and lowest and the difference between these, which is the temperature range. Quote figures from the graph to give added detail. Describe the temperature pattern in general terms, noting whether the climate is cold, cool, mild, warm, hot or very hot, and use these terms to describe the various seasons of the year.

◆ The graphs in Figure 1.10 show Tundra (cold desert), Mediterranean (warm temperate), Hot Desert and Equatorial climates.

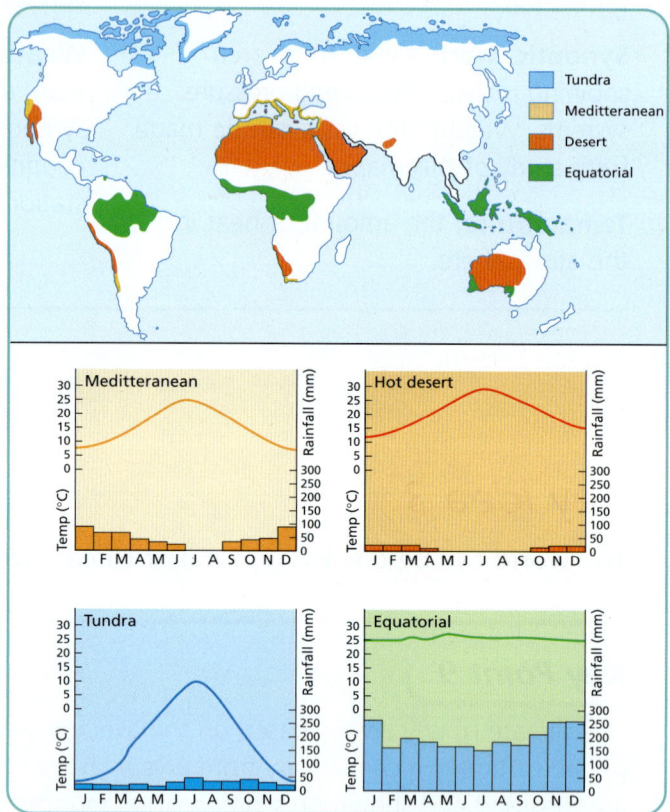

Figure 1.10 Four world climates

Key Point 11

You should be able to describe the above four climates in detail from given graphs.

Describing and analysing climate graphs

◆ Look at each of these climate graphs and describe them as outlined above (first bullet point).

◆ Normally marks for this type of description would range from 3 to 4 marks.

◆ Use the number of marks as an indication of the amount of detail needed.

◆ Remember to quote actual numbers and months from the graphs.

◆ Refer first to the temperature pattern, noting highest and lowest and the months.

◆ Refer secondly to rainfall, noting wettest and driest periods and giving amounts.

◆ Note the range of temperature – that is, the difference between the highest and lowest temperatures.

Sample Questions and Answers

Question: General level 2000

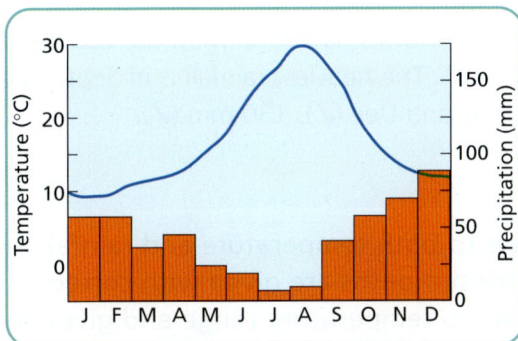

Figure 1.11 Climate graph for Athens (Greece)

Describe in detail the climate shown in Figure 1.11.   **(3 marks ES)**

Sample answer

In Athens the temperature is in contrast with the rainfall. When there is little rain the temperature is high (✔). When there is a fair amount of rain (J, F, O, N, D) the temperature stays around the same (✔).

Comments and marks obtained

This answer lacks sufficient detail for full marks. Extra marks could have been gained for referring to actual temperature and rainfall figures. As it stands it would be fortunate to gain **2 marks out of the 3** available.

Sample Questions and Answers continued ➢

Sample Questions and Answers continued

Question: General level 1999

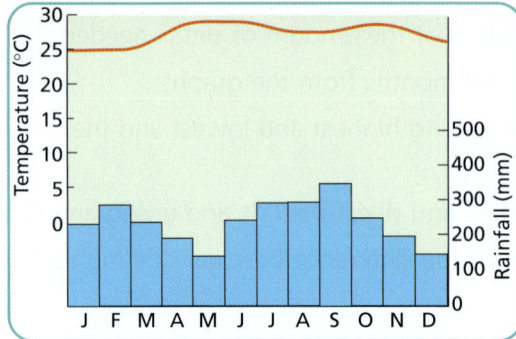

Figure 1.12 Climate graph of tropical rainforest

Look at Figure 1.12.

Describe in detail the climate of the tropical rainforest.　**(3 marks ES)**

Sample answer

There is little change in temperature (✓). The hottest is 29° in May, the coolest is 25° in Dec, Jan and Feb (✓). The range is 4°. The heaviest rainfall is in September (✓), 350 mm (✓), the least rainfall is in May and Dec (✓), 150 mm (✓).

Comments and marks obtained

This is a very good answer which refers to both temperature and rainfall in detail. Hottest, coolest, wettest and driest months are given with correct figures. The answer also correctly refers to temperature range and gives an accurate figure. The answer would have achieved at least a total of 6 marks had those marks been available. It does achieve a full **3 marks out of 3**.

Glossary of Associated Terms
Key Idea 3: World Climate Zones

Climate: the average conditions of weather usually taken over a period of 35 years.

Desert: any area which has very low rainfall throughout the year. Note that there are cold and hot deserts in the world.

Drought: this occurs when there is a long period without rainfall. Notice that droughts do not necessarily just happen in desert areas.

Latitude: the distance between the Equator and the Poles. It is measured in degrees.

Precipitation: all forms of moisture in the atmosphere including rain, hail, sleet and snow.

Rainfall pattern: this refers to the distribution of rainfall throughout the year. It may occur mainly at certain times of the year, e.g. winter or summer, or be spread evenly through the year.

Range of temperature: the difference between the highest temperature and the lowest temperature in a year.

Tundra: the cold desert area found only in the northern hemisphere between the upper limit of tree growth and the area of permanent snow and ice.

Key Idea 4

The physical environment offers a range of possibilities for and limitations on human activities.

Key Point 12

The type and range of human activities present in any area depend largely on the nature of the physical environment. Physical factors such as climate, relief, soil type, vegetation and presence or absence of a source of water greatly influence land use.

Key Point 13

You should know that given the appropriate combination of physical and human factors, many different land uses, such as industry, farming, settlement, communications, forestry, recreation and leisure activities, may be attracted to certain areas.

You should also know that a different combination of these factors may have the effect of actually repelling various land uses.

Examination questions based on this Key Idea often ask you to discuss the relationship between physical factors and land use. These questions are often based on an Ordnance Survey map on the scale of either 1 : 50 000 or 1 : 25 000. You should be able to identify various land uses on the map provided and then be able to discuss the impact of the physical landscape on their presence.

Land Uses and the Physical and Human Factors Which Affect Them

Industry

◆ Industrial land use can vary from large heavy industry to light industrial estates or may include power stations such as hydro-electric power schemes.

◆ The sites chosen for each of these industrial complexes are greatly influenced by factors such as the height and shape of the land, whether there is a water supply nearby, or the nature of the underlying rock strata.

◆ You may be shown certain environments on either a diagram, a map or an Ordnance Survey map, and be asked to determine the main factors which influence industrial land use.

Farming

◆ There are several different types of farming and each type has a close relationship with the physical environment.

◆ Climate often attracts or limits the type of farming present.

◆ Crops may flourish or die if there is plenty of rainfall, too much or not enough. Temperature throughout the year determines growing seasons.

◆ Farmers may opt for livestock farming as opposed to crop farming due to the influence of climate.

◆ The flatness of land or steepness of slope and the relative fertility of the underlying soil can encourage certain types of farms or limit others.

Settlement

◆ All settlements require access to water, flat land to build on and reasonably good access to other areas.

◆ Therefore landscapes such as valleys and plains are more attractive than hills, mountains and deserts.

◆ Specific site factors are dealt with in detail in the section under Key Idea 7, page 65.

Communications

◆ Accessibility is greatly affected by certain physical obstacles such as rivers, highland, marshland and climatic conditions.

◆ Engineers building routeways such as roads and railways must consider these factors and also how to overcome any difficulties presented by physical factors.

◆ Therefore, in General and Credit questions you may be asked to identify physical problems and offer certain solutions such as bridges, tunnels or embankments which may be used in the construction of communications.

Water storage schemes

◆ The presence of a reservoir, water storage scheme or hydro-electric scheme depends on physical factors such as the underlying geology, for example, it needs to be built on an impermeable rock such as granite.

◆ These schemes are often found in areas of high annual rainfall, such as the north-west highlands of Scotland, the Lake District and Snowdonia.

◆ Corrie lochs in Scotland and tarns in England provide suitable landscape features as sites for such land uses.

◆ Highland landscapes offer the differences in height that are needed for hydro-electric schemes. The water falls from a high level to drive turbines built at a lower level.

◆ Often such areas may be of little use for anything else such as farming, industry or settlement due to the difficulty of the terrain.

◆ Lochs and lakes offer natural reservoirs and are often used to supply water to large settlements.

Tourism, recreation and leisure

◆ Tourism is a major industry in most countries throughout the world.

◆ Many countries with warm and dry climates, or cold and snowy climates, may take advantage of their climate to build centres to attract tourists, such as coastal resorts or ski centres.

◆ Recreational pursuits such as golfing; sporting activities such as sailing, hill walking, and climbing; sightseeing and observing nature are closely linked to the nature of the physical environment.

◆ You have to be able to use the resources given in questions to draw conclusions on the influence of certain physical factors on specific human activities.

◆ Often much of this is based on common sense or perhaps on your skill in map interpretation.

◆ In many questions based on this topic there may not be a right or wrong answer, but the marks you gain will depend on how well you can relate individual factors to specific land uses from a given resource.

Sample Questions and Answers ?

Question: General level 1998

Figure 1.13 Glen Chanter estate in the Scottish Highlands

Look at Figure 1.13.

The new owner of Glen Chanter is considering four possible uses for his land.

These possible uses are to:

1   keep the land for deer stalking and grouse shooting
2   plant the area with trees
3   flood the valley for a reservoir and HEP station
4   make the area a Nature Reserve.

Choose one of the possible land uses and give arguments for and against it being the best use for the estate.   **(4 marks ES)**

Sample answer

Choice: Make a HEP station

For: A hydro-electric power station could be very good for the area. It will produce renewable energy sources (✔) for the people there meaning it won't cost as much for electricity (✔). It will bring many jobs to the area (✔).

Against: A hydro-electric power station would spoil the valley completely (✔). It would kill off animals because the valley would have to be flooded (✔). If the scheme doesn't work the whole area would be ruined because of the ugly-looking station in the valley.

Comments and marks obtained

The answer gains 3 marks for the three statements relating to the arguments 'for'.

It gains a further 2 marks for the reference to spoiling the valley through flooding and killing the wildlife. The final statement gets no marks since it speculates on how the station might look.

The answer gains **4 marks out of 4**.

Question: Credit level 2000

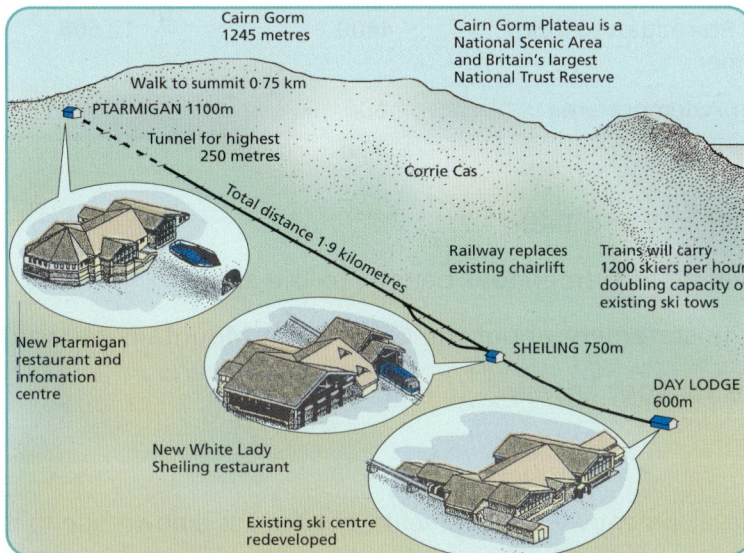

Cairn Gorm
1245 metres

Cairn Gorm Plateau is a
National Scenic Area
and Britain's largest
National Trust Reserve

Walk to summit 0·75 km

PTARMIGAN 1100m

Tunnel for highest
250 metres

Corrie Cas

Total distance 1·9 kilometres

Railway replaces
existing chairlift

Trains will carry
1200 skiers per hour,
doubling capacity of
existing ski tows

SHEILING 750m

New Ptarmigan
restaurant and
infomation
centre

DAY LODGE
600m

New White Lady
Sheiling restaurant

Existing ski centre
redeveloped

Figure 1.14 The Cairn Gorm Mountain Railway

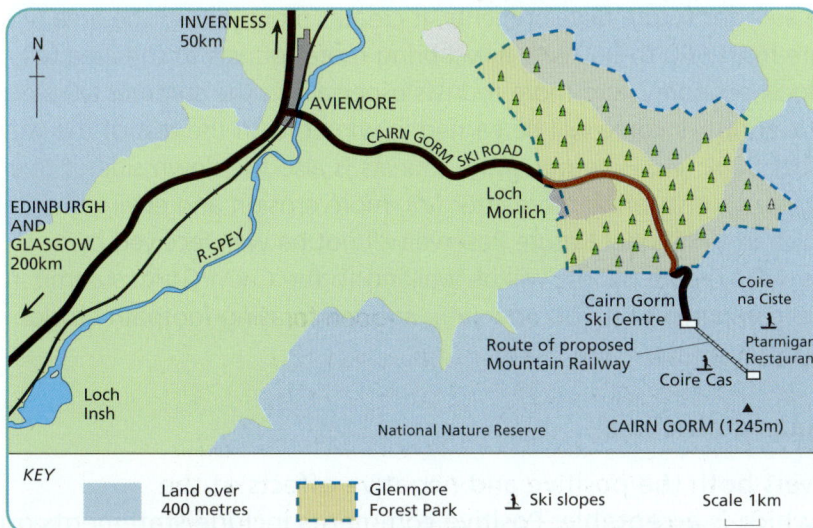

INVERNESS
50km

N

AVIEMORE

CAIRN GORM SKI ROAD

Loch
Morlich

EDINBURGH
AND
GLASGOW
200km

R. SPEY

Coire
na Ciste

Cairn Gorm
Ski Centre

Ptarmigan
Restaurant

Route of proposed
Mountain Railway

Coire Cas

Loch
Insh

National Nature Reserve

CAIRN GORM (1245m)

KEY

Land over
400 metres

Glenmore
Forest Park

Ski slopes

Scale 1km

Figure 1.15 Area around Aviemore and the Cairngorms

Sample Questions and Answers continued

Table 1.2 Selected statistics for Aviemore and the Cairngorms

	With existing ski tows and chairlifts	Estimated figures after opening of Mountain Railway
Cairn Gorm Ski Centre income	£3 million	£5 million
% of total income earned in winter months	90	50
People travelling up to Ptarmigan Restaurant in summer	55,000	125,000
People walking from Ptarmigan to Cairn Gorm summit in summer	4400	12,500
Tourist-related jobs in Aviemore area	600	960

Study Figures 1.14 and 1.15 and Table 1.2.

'This development will be of tremendous benefit to the area.'

Do you agree with this statement about the Cairn Gorm Mountain Railway?

Give detailed reasons for your answer.   **(6 marks ES)**

Sample answer

I do agree that the development will bring benefits, but there is a down side to it as well. It raises the income from £3 million to £5 million (✓), bringing lots more money into the area for future developments. It creates more jobs in the area (✓) raising the figure from 600 to 960 (✓). It will bring more tourists to the area (✓), increasing the local economy, and more tourists especially in the summer (✓) since 50% of the total annual income will be earned in winter, with the rest of the year bringing the other 50% (✓). However, more tourists is also the down side of this development. More tourists bring more litter (✓), more erosion and environmental damage (✓) which in a National Nature Reserve will not be well received by conservationists. Many more people will be walking to the Cairn Gorm summit in summer (12,500 compared to 4,400 now (✓)), again increasing footpath erosion (✓) and possibly litter up the Cairn Gorm.

Comments and marks obtained

This answer covers both the positive and negative effects of the development, which is acceptable. Positive comments include statements on income, jobs, and extra tourists in summer, with examples, all worth a total of 6 marks. Comments on the problems achieve a further 4 marks for references to litter, environmental damage, increased numbers and footpath erosion.

Overall the answer has more than sufficient correct statements to achieve the full **6 marks out of 6**. The candidate makes very good use of the resource material without simply resorting to straight lifting of information from the diagrams.

Glossary of Associated Terms
Key Idea 4: The Physical Environment and Human Activities

Country Park: area in the countryside surrounding a town or city which has been set aside for people to visit as a park.

Countryside Commission: organisation set up by the government to monitor and protect countryside areas from harmful development.

Forestry Commission: organisation responsible for planting and looking after forests throughout the UK.

Hydro-electric power schemes: schemes which often take advantage of the physical landscape to generate electricity using water to power turbines. Often these schemes use off-peak electricity to pump water back up to the reservoir to ensure a steady supply of water during periods of low rainfall.

Land use: how humans make use of the physical landscape, e.g. forestry, farming, industry or settlement.

Land use conflict: occurs when different activities compete with each other to make use of the land, e.g. farming and tourism.

National Park Authority: organisation which looks after areas that have been set aside throughout Britain for the recreation and enjoyment of the public. Its aims also include protecting these areas of outstanding scenic beauty.

Reservoirs: bodies of water used for water supply. They may occur naturally, such as those in glaciated areas, or they may be created artificially by building dams either across rivers (flooding valleys) or across the lip of corries and tarns.

Key Idea 5

There are many competing demands for the use of rural landscapes.

Key Point 14

Figure 1.16 summarises some of the main competing land uses in rural (countryside) areas. This topic is examined only within the context of Scotland.

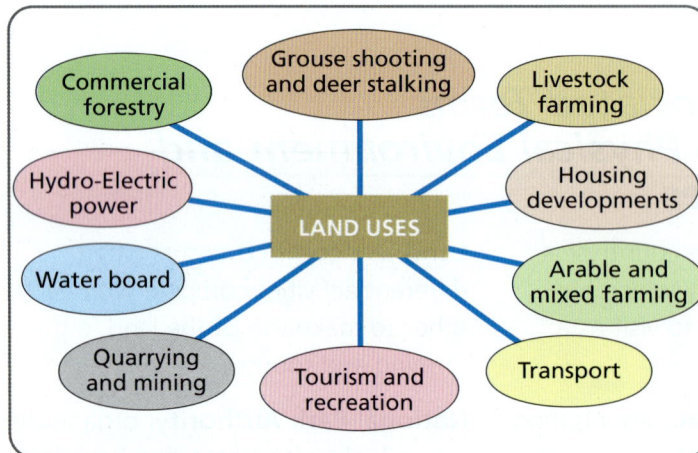

Figure 1.16 Competition for land use

Land Use Conflicts: Points for Consideration

◆ At all levels you can be asked to describe the reasons why certain land uses are in conflict with each other. Usually you will be asked to select two or more from a given list or diagram.

◆ Alternatively you may be asked to look at an Ordnance Survey map and suggest land uses which may be in conflict and to account for the nature of the conflict.

◆ In rural areas most other land uses are in conflict with farming.

◆ **Farming** needs land, and all other land uses such as settlement, industry, forestry or reservoirs obviously limit the amount of land available. Land use which produces any kind of pollution, such as fumes, litter or chemicals, can have a detrimental effect on crops and livestock. Therefore, industry, settlement, communications and tourism are often in direct conflict with agriculture for this reason.

◆ **Tourists** in many rural areas can seriously affect other land uses either directly or indirectly. Large numbers of tourists require certain facilities such as roads and other services. Traffic issues such as road congestion and vehicle fumes can affect natural vegetation and disturb the peace and quiet of the rural environment.

◆ **Industrial** and **settlement** land uses take up large areas of the countryside. Cities are always growing outwards and often encompass rural areas, including small rural settlements.

Key Point 15

Not only should you be able to identify when and where conflicts exist, but you should also be able to describe the kinds of measures which can be taken to resolve these conflicts, both at local and national levels.

◆ There are certain government agencies, at local and national levels, whose task it is to monitor and, wherever possible, prevent or solve major conflicts.

◆ National Park Authorities, Country Park Authorities and local pressure groups often ensure that the countryside is protected and that conflicts caused by competing land uses are resolved.

◆ Occasionally this can involve legislation or restricted access to certain areas, or protective measures such as constructing footpaths to guard against erosion.

◆ Some questions may involve looking at different points of view. You need to make use of these in your answers to support certain proposals, for example, the building of a new by-pass route or outdoor centre or the construction of a new industrial site.

Sample Questions and Answers

Question: General level 2000

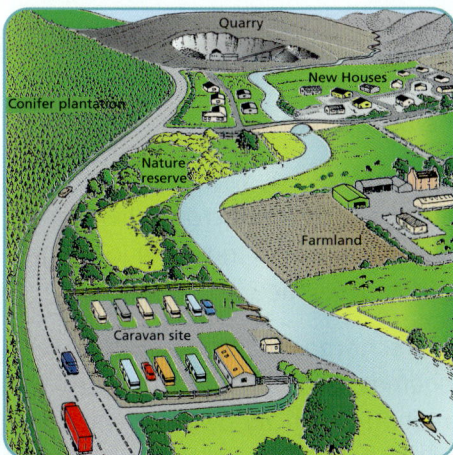

Figure 1.17 Land uses in the countryside

Sample Questions and **Answers** continued ➤

Sample Questions and Answers continued

Look at Figure 1.17.

'There are possible conflicts between the land uses shown.'

Do you agree with this statement? Tick the box ☐ Yes ☐ No

Explain your answer.   **(4 marks ES)**

Sample answer

The farmer may get annoyed with the people from the caravan site (✓) because people could go and drop litter (✓) on the ground or in the river and then it may end up on his land. The people in the new houses may get annoyed with all the heavy machinery (✓) from the quarry making lots of noise (✓).

Comments and marks obtained

This is a good answer which discusses the conflict between the farmer and those using the caravan site and local industry. It correctly identifies three possible sources of conflict, namely litter, heavy machinery and noise from the quarry. The answer deserves the full **4 marks out of 4**.

Glossary of Associated Terms
Key Idea 5: Competing Land Uses

Conflict: this happens when two or more land users disagree as to what is the best use of the land. Often one land use is totally at odds with another, such as industrial activities spoiling and polluting areas used for farming.

Conservation: efforts made to maintain the basic beauty and attractiveness of areas both in the countryside and in towns.

Countryside: the majority of land area which has not been used for towns or cities.

Countryside Commission: government agency set up to protect and monitor the countryside in Britain.

Habitats: places where animals and birds live, e.g. hedgerows, fields and woodland. Often these are lost when land is developed.

HEP schemes: hydro-electric power schemes, built for the purpose of creating electricity by using water in reservoirs as a source of power to drive turbines. The electricity is fed into the national electricity grid.

Honeypots: places which are highly attractive to tourists and often very busy, especially during the holiday seasons.

Irrigation: system of bringing water to dry areas.

National Parks: places of outstanding scenery set aside and protected by legislation which is enforced by the National Park Authorities. They are set up for the purpose of attracting people from urban areas for leisure and recreational purposes. There are eleven National Parks in England and Wales – Northumberland, North Yorkshire Moors, Yorkshire Dales, Lake District, Peak District, Snowdonia, Pembrokeshire Coast, Brecon Beacons, Exmoor, Dartmoor, Norfolk Broads and two in Scotland – Loch Lomond and the Trossachs, and the Cairngorms.

Rural: another name for countryside areas.

Settlement: places where people live and work, ranging from small villages to large towns and cities.

Key Idea 6

The physical environment is a resource which has to be used with care. Its management is a global issue.

Key Point 16

Questions on this part of the syllabus concentrate on three main topics:

◆ the deforestation of tropical forests
◆ the process of desertification
◆ the threat to the world's oceans and seas due to use and misuse.

Essentially these questions will refer either to the causes and/or to the effects and/or to the solutions to these threats to the natural environment.

◆ Questions may be based on your specific knowledge of these topics (KU).
◆ Alternatively, they may be based on a resource from which you can form an opinion, using the resource to support your answer.
◆ This type of question may ask you to consider the advantages or disadvantages of the processes creating the problems, or the advantages and disadvantages of the measures taken to reduce the effects.

Deforestation (Tropical Rainforests)

Throughout equatorial areas of the world thousands of square kilometres of trees have been destroyed in various ways and for a variety of reasons.

Methods of destruction

- Destruction by fire, which may be the easiest and quickest way to destroy forests.
- Trees being cut down by logging workers, since hardwoods such as teak and mahogany can be sold for large sums of money abroad.
- Trees drowned by valleys being flooded to create reservoirs for multi-purpose water schemes.
- Despite the damaging effects on the environment, people in countries such as Brazil and the Congo and parts of India and South East Asia continue to destroy the forest.

Reasons for the destruction

- Cutting down hardwood trees for timber.
- Clearing areas for small-scale farmers.
- Clearing trees for mining of a variety of minerals such as iron ore, copper, bauxite and gold using, for example, high-powered water jets.
- Using trees for fuel, for example, charcoal for local industry such as iron smelting.
- Clearing large areas of forest to create grazing land for cattle ranching.
- Flooding valleys to create reservoirs for hydro-electric power schemes.

All of these reasons are basically aimed at obtaining money from destructive activities.

The effects of deforestation

- Destruction of trees can have a devastating effect on people, their homes and their livelihood.
- Thousands of different species of animals and insects are destroyed each year.
- Thousands of different species of plants are also destroyed each year.
- The Earth's atmosphere is seriously affected by the release of carbon dioxide and the loss of oxygen.
- Deforestation is known to contribute to global warming and the greenhouse effect, leading to a rise in sea levels throughout the world.
- Rivers become polluted from mining enterprises, killing fish and affecting food supplies and the welfare of tribes living in the forest.
- Without trees to interrupt run-off of rainwater, serious flooding occurs in other areas causing widespread disaster and thousands of deaths.

Measures to reduce deforestation

- Governments pass laws to protect forests by limiting the amount of land which can be used for activities such as mining and ranching. These laws are often very difficult to enforce.

- There are worldwide campaigns by various protest groups such as Greenpeace directed at governments to save the forests.
- Other commercial developments are encouraged within the forests by promotion of forest products such as tropical fruits.
- There are many who live in the forest who can work in harmony with the forest, including tribes who hunt and gather and who are subsistence farmers, rubber trappers, farmers, and those who replant felled areas with young trees.
- Unfortunately the process of deforestation is continuing faster than the efforts to reduce it.

Sample Questions and Answers

Question: Credit level 2000

Study Figure 1.18(i) and Table 1.3.

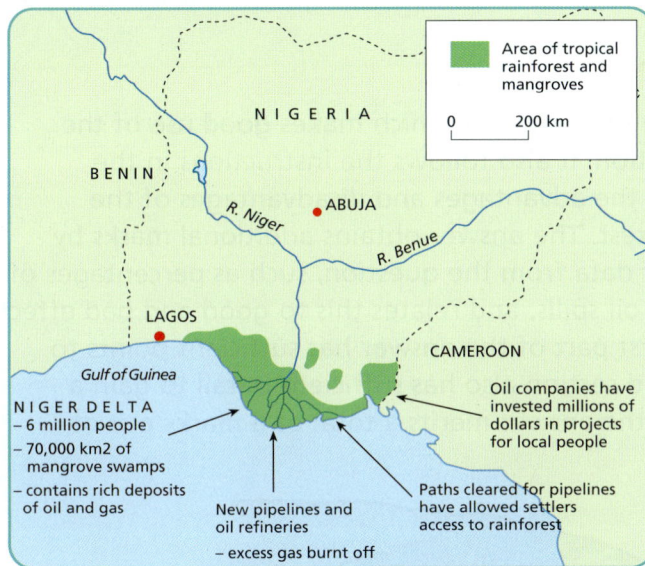

Figure 1.18(i) Oil industry in the Niger Delta (West Africa)

Table 1.3 Facts about oil in the Niger Delta

% of Nigeria's exports provided by oil	90
Number of oil spills in the Niger Delta 1976–1991	2976
Number of people employed in the oil industry in Nigeria	25,000

'The damage caused to the rainforest environment in the Niger Delta is a small price to pay for the huge benefits which the oil industry has brought to the people of the area.'

Do you agree with this statement? Give reasons for your answer. **(6 marks ES)**

Sample Questions and Answers continued (?)

Sample answer

I agree that there are huge benefits from the oil industry but the rainforests should not have to be cut down. The oil companies have invested lots of money in projects for local people which gives jobs (✔) and therefore money to buy food and clothes and proper houses (✔). This is an excellent idea for these people. The number of people employed in the oil industry is 25,000 which is great for the country (✔). The oil provides 90% of Nigeria's exports as well (✔).

Despite all of this the rainforests are being destroyed because of it. Many people's houses and habitats are being destroyed with the forests (✔). The new pipelines are built and thousands of people and wildlife lose their habitats because of it.(✔). There were 2,976 oil spills between 1976 and 1991 and nobody knows the amount of wildlife this has killed (✔). The trees provide the world with vital vitamins and minerals and other chemicals which are used to make medicines (✔).

! Comments and marks obtained

This is a very good example of an answer which makes good use of the resources given in the question. It also follows the instructions in the question, namely to discuss the advantages and disadvantages of the developments in the rainforest. The answer obtains additional marks by including relevant pieces of data from the question, such as percentages of exports and the number of oil spills, and relates this to good and bad effects of the development. The first part of the answer has sufficient points to gain 4 marks and the second section also has sufficient detail to gain a further 4 marks. Therefore the answer merits a total of **6 marks out of 6**.

Sample Questions and Answers continued

Question: General level 2004

Look at Figure 1.18(ii)

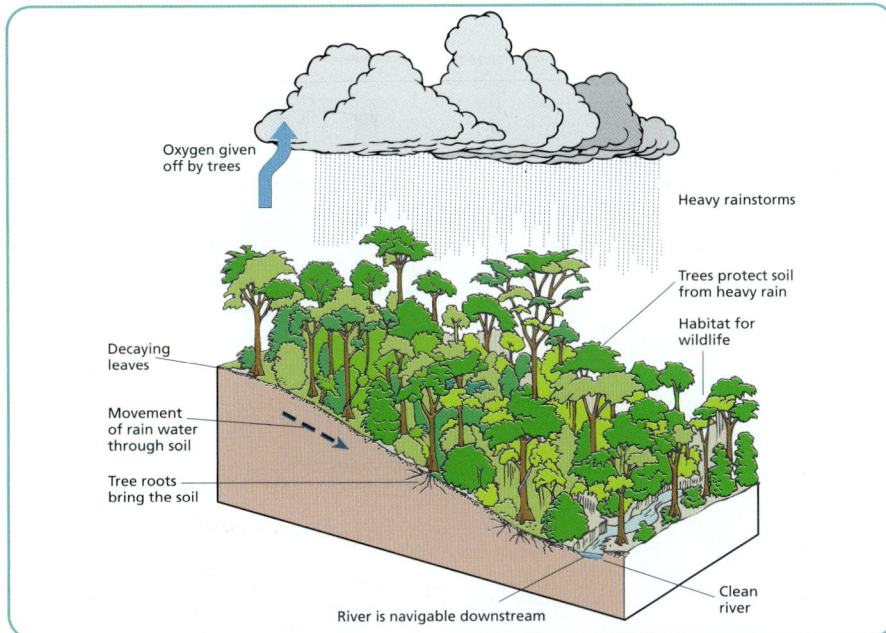

Oxygen given off by trees

Heavy rainstorms

Trees protect soil from heavy rain

Habitat for wildlife

Decaying leaves

Movement of rain water through soil

Tree roots bring the soil

Clean river

River is navigable downstream

Figure 1.18(ii) A tropical rainforest landscape

Explain problems caused by deforestation in the Tropical Rainforest.
(4 marks KU)

Sample answer

If the trees are taken away there will be nothing to protect the soil (✓). The tree roots will no longer be binding the soil together so it will be loose (✓). Heavy rainstorms will erode away the soil and wash it down into the river (✓). The water will then be dirty (✓) and animals will not be able to drink from it and they will not have anywhere to live (✓) if there are no trees, or have anything to eat.

Comments and marks obtained

This answer has five good points relating to the problems of deforestation. Marks are obtained for the references to lack of soil protection by removal of the trees, the effect on the soil by the removal of tree roots, soil erosion caused by heavy rain, pollution of the river and the loss of habitat for wildlife. The lack of drinking water and loss of food could possibly have gained a further mark if it was needed. The answer gains full marks – **4 out of 4.**

Sample Questions and Answers continued

Question: Credit level 2004

What measures could be taken to reduce the impact of the threats to the marine environment? (as shown in Figure 1.18 (iii)) **(6 marks KU)**

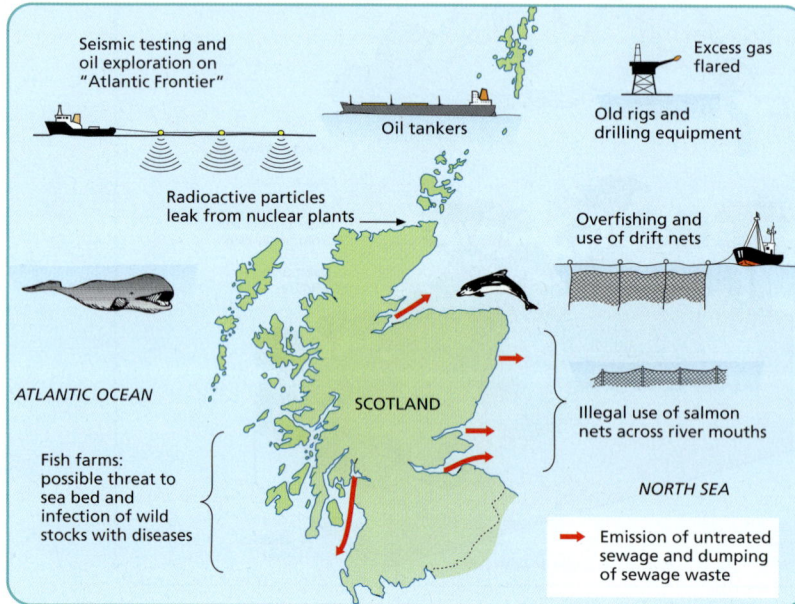

Seismic testing and oil exploration on "Atlantic Frontier"

Oil tankers

Excess gas flared

Old rigs and drilling equipment

Radioactive particles leak from nuclear plants

Overfishing and use of drift nets

ATLANTIC OCEAN

SCOTLAND

Illegal use of salmon nets across river mouths

NORTH SEA

Fish farms: possible threat to sea bed and infection of wild stocks with diseases

Emission of untreated sewage and dumping of sewage waste

Figure 1.18(iii) Threats to the marine environment of Scotland

Sample answer

Less pollution could be allowed into the sea to stop the threat on marine life. Increased restrictions could be put on fishing (✔) and a 'clamp down programme' could be introduced to combat the salmon nets (✔). Restrictions could be put on fish farms to reduce the risk of disease. Oil tankers could be limited and made safer with safety checks (✔) etc. Nuclear power stations could be made safer by increased protection to stop radioactive particles getting into the water. Oil rigs could barrel excess gas rather than burn it.

Comments and marks obtained

Although this answer would appear to cover enough points for full marks the answer has several faults. The opening sentence is too vague to gain a mark. If it had been more specific by referring to actual measures to prevent pollution, for example, reference to legislation, it could have been given a mark. Whilst it gained marks for reference to increased restriction on fishing and clamp down programmes on salmon nets, the candidate makes the error of repeating these points in the next sentence. This is a common fault in many answers. The specific reference to safety checks on oil tankers merits a further mark. However, the lack of specific detail in the sentence on nuclear power stations and reference to using barrels for excess gas is confused and consequently no further marks are obtained.

Many candidates answering this kind of question lose marks for being vague in reference to measures taken, or by saying 'measures' without specifying the measures which could be taken to 'stop' this and 'stop' that situation. Also, marks are lost by candidates repeating the same measure or by simply lifting data from the resource without further elaboration.

By doing this, despite mentioning several points, at Credit level, this answer only obtains a total of **3 marks out of a possible 6**.

Desertification

Throughout the world, the hot deserts are spreading at their edges into areas which were formerly settled and provided a living for many people. Although some physical factors are responsible for this process, including prolonged periods of drought, perhaps the agent most responsible is people. The impact of many human activities has led to land turning into desert.

Causes of desertification

◆ The main physical cause of desertification is a prolonged absence of rainfall, in other words drought.

◆ As a result plants and vegetation die and therefore there is little cover over the soils.

◆ A second factor is the effect of wind which blows away dead vegetation and topsoil, eroding some areas and depositing fine sand elsewhere.

◆ Insects may also contribute by eating vegetation and helping to remove soil protection.

◆ Human activities help to increase the process. These activities include, for example:

  – deforestation, by which trees may be cut down for firewood

  – farmers allowing their animals, such as sheep and goats, to overgraze the vegetation leading to more soil erosion

  – farmers over-cultivating the land and removing any goodness in the soil so that the soil cannot support any more crops.

◆ People do this because they have a desperate need for food during periods of extensive drought, and because there is a lack of expertise in proper farming techniques.

Effects of desertification

◆ Since the land cannot sustain any further farming, local people have little option but to move away to areas where the soil might be more fertile.

◆ Villages are abandoned.

◆ Without anyone to care for the land, land which was formerly fertile and settled gradually turns to desert, thus increasing the extent of deserts.

Efforts to reduce desertification

◆ The most effective measure to reduce this process is to bring water to dry areas.

◆ This can be done through various forms of irrigation, especially digging wells and ditches, or through large-scale multi-purpose water projects to bring much needed water to dried-up land.

◆ Governments can help farmers by advising them on better farming methods such as using fertilisers and 'miracle seeds' where possible.

◆ Grassland areas can be fenced off to avoid overgrazing.

◆ Soil conservation methods can be used.

◆ Young trees may be planted to act as windbreaks.

◆ Any of the actions which lead to the speeding-up of the process of desertification should be avoided.

◆ Much-needed financial assistance is provided to governments through international aid agencies such as the World Health Organisation, United Nations Food and Agricultural Organisation and the World Bank. Loans and aid programmes from more developed countries throughout the world can also help.

At both General and Credit levels you can be asked in *Knowledge and Understanding* questions to discuss the main causes, effects and methods to control the processes of deforestation and desertification.

Enquiry Skills questions:

◆ If you are asked for an opinion on these, or to judge the main advantages or disadvantages of different methods of controlling the problem, you will be provided with a resource such as a map or diagram.

◆ You must be able to use the information given in the source to support any points which you make in your answer.

◆ Remember: there is not always a right or wrong answer to Enquiry Skills questions.

◆ Marks are gained for selecting the appropriate data in support of your opinions or choice.

- The main difference in your answers between General and Credit levels is the amount of detail which you provide in your answer.

- Use the number of marks allotted to the question as a guide to the depth and amount of detail to give.

Sample Questions and Answers

Question: General level 2002

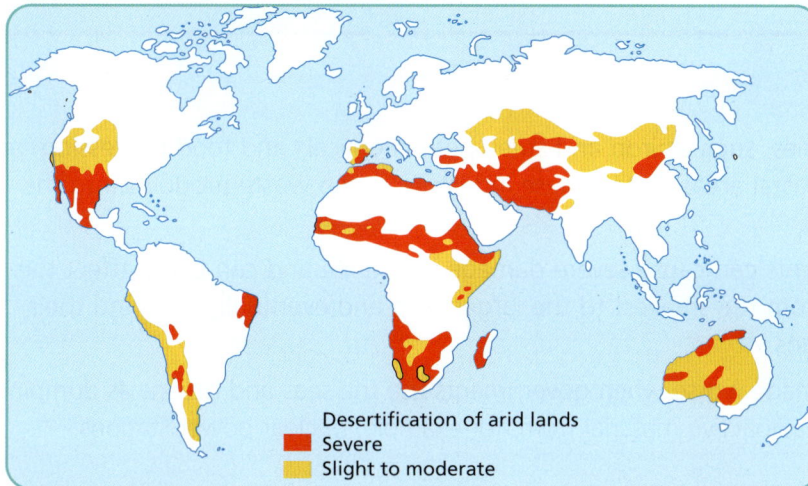

Figure 1.19 Desertification

Study Figure 1.19.

What are the main causes of desertification?   **(4 marks KU)**

Sample answer

When you clear away trees, there are no longer any roots to hold the topsoil together (✔) so this blows away causing desertification (✔). Over-farming – not allowing any fields to lie fallow (✔) and constantly grazing animals (✔) and growing crops on the same land every year – drains the soil of nutrients (✔) so again it blows away, causing desertification.

Comments and marks obtained

Although this answer does not refer to some of the main causes of desertification such as prolonged periods without rain, i.e. drought, nevertheless by referring to tree removal and its effects and to over-farming of the land, giving good examples of how this happens, the answer does gain enough marks to deserve more than the 4 marks available. It therefore gains **4 marks out of 4**.

Use and Misuse of the World's Oceans and Seas

The third area which is examined within this Key Idea is that of the use and misuse of oceans and sea areas such as the Mediterranean and North Sea.

Key Point 17

You should know the main causes of misuse of major water areas. Essentially this involves widespread pollution from various sources.

Industrial

◆ Many industries, such as iron and steel, petrochemicals and food processing, are located in coastal areas. Often these industries dump waste products into the sea or ocean.

◆ These pollutants can cause severe damage to sea life and can badly affect the food chain from the smallest to the largest fish and eventually may find their way to humans.

◆ Matters are made worse when governments use the seas and oceans as dumping grounds for radioactive material from, for example, nuclear power stations.

Transport

◆ All sea-going craft can add pollutants by dumping refuse and other garbage into the sea during their journeys. Given time this material can disintegrate but its sheer volume is causing a hazard in both seas and oceans.

◆ Oil spillages from ocean-going tankers are a major problem.

◆ Leakages from ships, especially when cargoes are being discharged, can gradually build up into a major misuse of sea areas.

◆ Worst-case scenarios involve major accidents in which large volumes of oil may be spilled causing widespread damage to fish, birds, beaches and coastal settlements. Such damage may take years to repair.

Domestic

◆ Sewage has to go somewhere. Some of it may be put into landfill sites or burnt off.

◆ A great deal of it finds its way through sewage pipelines from major cities into sea areas.

◆ These pipelines may extend several kilometres out into the sea, but even so they can affect local sea life, and much of this sewage may eventually, through the work of tides, find its way back to the coast. This type of beach pollution is very common.

Farming
◆ Farmers spread natural and chemical fertilisers and insecticides onto their crops.
◆ These chemicals can eventually find their way into rivers, polluting the waters which discharge into the sea.
◆ Seas and oceans are in great danger from modern methods of farming.
◆ These chemicals affect oxygen levels within the water and as a result fish die.

Tourism
◆ Many sea areas have coastal resorts catering for tourists, especially in regions such as the Mediterranean. The tourist industry can be an immense source of pollution of seas and beaches.
◆ Rubbish left on beaches by tourists can find its way into the sea.
◆ Tourist traffic giving off fumes can pollute the air causing acid rain.
◆ Sewage disposal is increased by tourists.
◆ Farmers may increase pollution by trying to improve yields to meet the demand of tourists for food supplies.
◆ Tourists mean increased demand for transport whether by road, sea or air, and this can all add to pollution.

All of the above points may be examined in a variety of questions at all levels. Those that test Enquiry Skills by asking for opinions will be based on an evaluation of a given resource.

Use the material within the resource in your answer but do not rely on simply copying the information. You must include original comments based on your assessment of the resource material. Direct quotations from the resource without any comments from you may not be sufficient to gain full marks.

Methods of Controlling Misuse
◆ Government legislation can be put in place to protect areas at risk, such as creating Environmentally Sensitive Areas, and coastal areas being developed for tourists.
◆ International agreements between governments can enforce protection legislation, for example, the Mediterranean 'Blue Plan'.
◆ Various national and international agencies have been set up to monitor and protect the environment of coastal areas. These include unofficial bodies which organise protests over misuse such as the building of power stations which could cause pollution.
◆ Tourist agencies can control access to certain areas and put in place measures to clean beaches.

◆ Rubbish disposal can be improved, for example, extensive use of litter bins and active clean-up operations.

◆ Education and advertising programmes help tourists become aware of the dangers.

◆ Various plans have been put into effect to clean up affected areas such as those to monitor the pollution levels and cleanliness of Britain's beaches. Similar measures have been adopted in many other countries.

◆ Some governments have made laws to protect sea environments. These allow prosecutions of the major causes of pollution and misuse such as tanker companies, industrial companies, and city sewage disposal departments.

◆ It is difficult to enforce these measures and the process can involve many years of court cases being heard before successful prosecutions are enforced.

Many of the questions relating to this topic involve discussing the advantages of different methods of control and their effects.

Sample Questions and Answers

Question: Credit level 1999

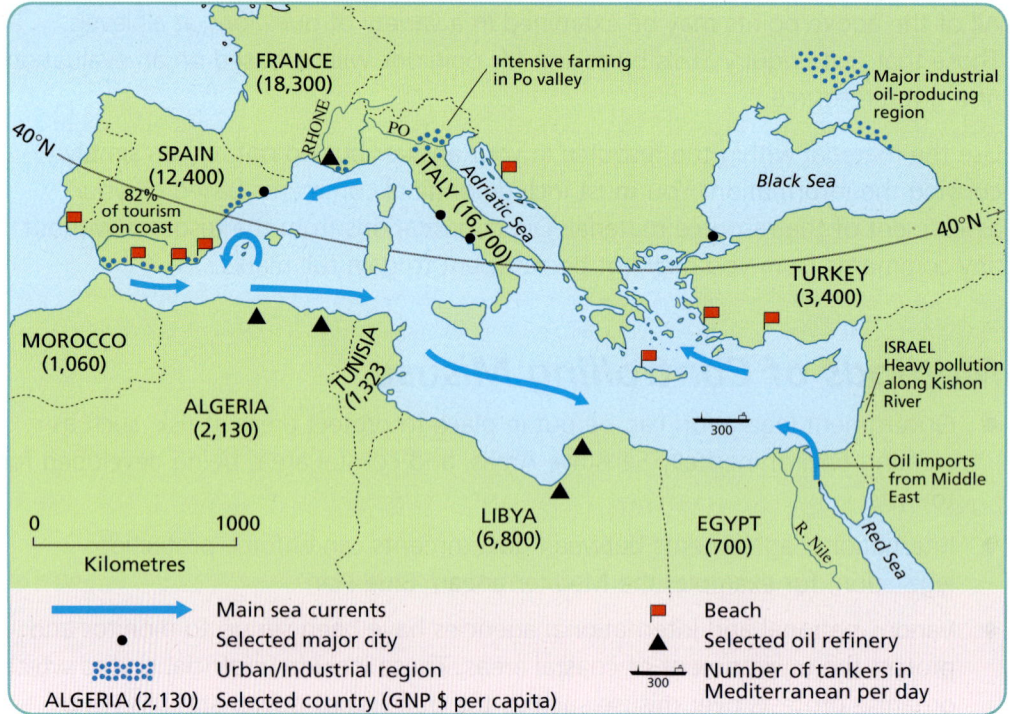

Figure 1.20 Pollution in the Mediterranean

Sample Questions and Answers continued ➢

Sample Questions and Answers continued

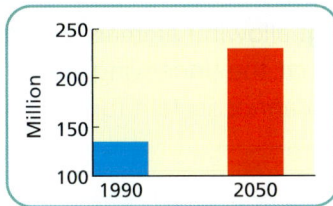

Figure 1.21 Expected population growth of the Mediterranean coastal regions

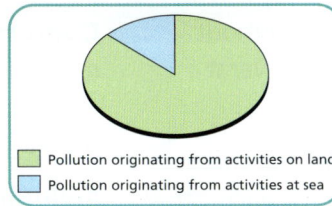

Figure 1.22 Sources of pollution in the Mediterranean

Look at Figure 1.20, 1.21 and 1.22.

Explain why the Mediterranean Sea suffers from severe pollution.
(6 marks ES)

Sample answer

There are 300 oil tankers in the sea every day. If they leak oil they can pollute the sea (✓). All the rubbish from ships gets thrown into the sea (✓). Chemicals from coastal oil refineries can also be dumped (✓). Rubbish from beaches can also get washed into the sea (✓).

Comments and marks obtained

Four correct sources of pollution are identified, each gaining mark. The candidate has failed to refer to pollution from farming, such as fertilisers and insecticides, eventually getting into the sea from rivers.

Nor has he/she developed any of their other points which could have gained further marks. The answer gains a total of **4 marks out of 6**.

Glossary of Associated Terms
Key Idea 6: Management of the Physical Environment

Blue Plan: an international agreement signed by countries which have a coastline on the Mediterranean Sea, to protect the sea and coastal areas from environmental pollution by sources such as industry, farming and tourism.

Cattle ranching: rearing of large herds of cattle on areas of cleared forest to provide beef to be sold for export.

Deforestation: removal of trees, usually on a large scale.

Desertification: process of land that was formerly productive turning into desert.

Drought: prolonged period without rainfall which may last from several months to several years.

Food chain: a system whereby various forms of life provide food for each other starting at one point and finishing at another, e.g. from small fish to larger fish and eventually to humans.

Global warming: the heating-up of the Earth's atmosphere by the sun's rays due to the effect of atmospheric pollution.

Greenhouse effect: a rise in temperature in the Earth's atmosphere due to the effect of increased carbon dioxide and other gases in the atmosphere from various sources such as industrial pollution, burning forests, car exhaust fumes and smoke from domestic chimneys.

Greenpeace: an international organisation which protests against the causes of world pollution and other elements which destroy the natural environment.

Insecticides: chemicals used by farmers to kill insects which feed on crops.

Irrigation: an artificial means of providing water for farming from sources such as rivers, wells, canals, field sprays.

Logging: a commercial business which cuts down trees to provide timber for sale for different purposes.

Overgrazing: allowing animals to eat so much grass that the underlying soil is exposed and cannot sustain further growth.

Pollutants: material which is released into the environment and which ultimately causes damage to the physical landscape and the atmosphere.

Power stations: complexes which are built to provide electricity from various sources, e.g. nuclear energy, fossil fuels and water.

Shelter belt: a line of trees planted to provide shelter from the wind for fields by interrupting the flow of wind.

Soil conservation: attempts to protect soil from damage using methods such as fertilisers, irrigation, shelter belts, ploughing along contours and creating terraces to conserve soil.

Soil erosion: the process by which the topsoil is removed leaving the land infertile.

Tundra: an area of cold desert located in the northern hemisphere. It lies between the northern limits of tree growth and the areas of permanent snow and ice.

Windbreaks: line of trees which interrupts the flow of wind so as to protect fields and their crops.

Chapter 2

THE HUMAN ENVIRONMENT

i

Key Idea 7

Settlements have many common characteristics related to site, situation and function.

> **Key Point 18**
>
> Settlements are places where people gather together to live and work. They vary in size from the smallest (isolated houses) to hamlets to villages to towns to cities.

- ◆ **Site** refers to the actual ground on which the settlement is built.
- ◆ **Situation** refers to the position of the settlement in relation to other landscape features and to other settlements.
- ◆ **Function** refers to the main purpose of the settlement, for example, a residential area, an industrial area, a place offering services such as shops and businesses for its residents and the local area, a port, a religious centre, an educational centre, a holiday resort – or perhaps a mixture of some or all of these.
- ◆ **Function** can also refer to specific features within a settlement, such as residential, industry, transport, education, medical, financial, public administration, retail, wholesale, commercial, recreational, religious, cultural.
- ◆ Each of these functions would be found in particular types of buildings such as shops, offices, libraries, council offices, churches, hospitals and clinics, schools and colleges, bus terminals and railway stations, factories or industrial estates, banks and estate agents, museums, cinemas and theatres, business premises, sports and recreational centres.
- ◆ The number and range of these will vary according to the size of the settlement.

Settlement Site

The original site of any settlement would have had certain features which attracted people to settle there in the first place. These features may have included:

◆ flat land for building on

◆ a source of water such as a nearby stream or river

◆ some form of defence such as the meander of a river or the top of a hill

◆ woodland located nearby for building materials

◆ a suitable narrow point on a river where it may be crossed by building a bridge

◆ the entrance to a gap between two hills

◆ local mineral resources which could be used to develop an industry such as mining.

Any one or a combination of these features may be important in the original building of a settlement. Most questions relating to this topic will be based on the examination of an Ordnance Survey map.

◆ You should look at the map to find clues to original site features.

◆ Contour patterns will indicate flatness or steepness of the land (relief) and whether it is situated in a gap between hills.

◆ Water sources such as rivers or streams can easily be seen on a map.

◆ Local resources may be determined from existing or closed mines and quarries.

◆ Bridges will indicate if the need to cross a river has been a factor.

◆ The number of factors you refer to will depend on the number of marks available for the question, regardless of the level.

◆ Questions relating to situation will usually be accompanied by resources such as a map or diagram.

◆ You should refer to communications and distances to other settlements within the area. Communication can be by road, rail or water such as rivers or canals.

◆ Mention appropriate grid references when referring to communications.

Functional Zones

In certain areas or zones of towns and cities, specific function types such as shops, businesses or housing may dominate.

These areas are called land use zones or sometimes 'functional zones'.

You should be able to describe the main features or landscapes of the different land use zones found within a city.

Figure 2.1 shows a summary of the main zones within cities. If these zones are looked at together with the land use values, you should be able to detect a direct link between them.

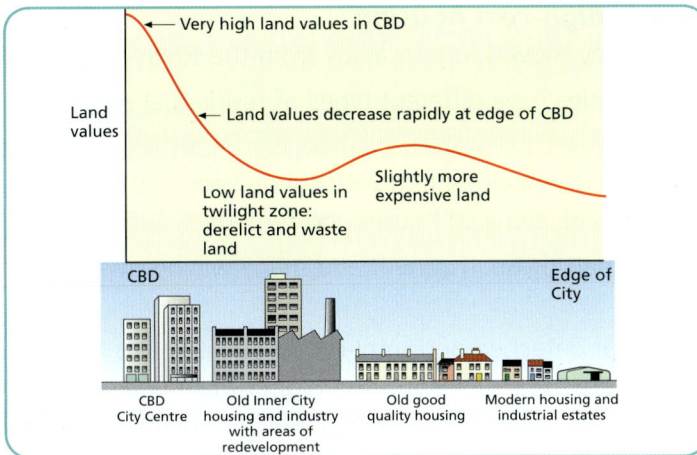

- Very high land values in CBD
- Land values decrease rapidly at edge of CBD
- Low land values in twilight zone: derelict and waste land
- Slightly more expensive land

Land values

CBD — Edge of City

CBD City Centre | Old Inner City housing and industry with areas of redevelopment | Old good quality housing | Modern housing and industrial estates

Figure 2.1 Land use zones

Features of the main functional zones in a city

Zone 1 The central business district

- ◆ Those functions which provide important services for residents of a settlement are usually located at or near the centre. This area is called the Central Business District (CBD).

- ◆ You may be asked to identify this area on an Ordnance Survey map by indicating certain map squares. You may also be asked to explain your choice.

- ◆ You should discuss the types of buildings, the pattern of streets (usually narrow or in a grid-iron arrangement), the fact that major roads converge there giving greater accessibility, and that there should be bus and railway stations and public buildings such as town halls there.

Zone 2 Old industrial zone

- ◆ Old industrial areas with factories (works) are usually located close to the CBD.

- ◆ Accessibility is normally the main reason for this, which allows factories to bring in their raw materials and distribute their finished products.

- ◆ It also allows workers to get to their workplace easily.

Zone 3 Low-cost housing

- ◆ Since the area around old factories is not the most pleasant to live in, housing around this area is usually the cheapest and would have originally been built for workers and their families.

- ◆ On an Ordnance Survey map the streets may form a grid-iron pattern, with narrow streets.

- ◆ The density of housing will usually be high, consisting mostly of tenements.

- ◆ Population density will also be high in these areas and many of the houses may be fairly old.

- ◆ Zones 2 and 3 make up the area known as the 'inner city'.

Zones 4 and 5 Medium and high-cost housing

◆ As people could afford to, they moved further away from the town centre.

◆ As a result most towns and cities have different types of residential areas: low, medium and high-cost areas.

◆ In zone 4, streets may have a curvilinear pattern with cul-de-sacs, and houses may consist of a mixture of tenements, terraced houses and semi-detached houses.

◆ House prices will be higher than those in zone 3 and many will be owner-occupied.

◆ Zone 5 may have a mixture of detached and semi-detached villas and bungalows which may be more expensive and have higher maintenance costs. The streets may be wider and the density of housing will be much less than in any other residential zone.

◆ On the fringes or outskirts of cities and towns there are areas which have been used for newer, more modern industries and out-of-town shopping areas.

◆ Some cities, especially in Scotland, have council house schemes with lower-cost housing built in these areas.

Zone 6 Commuter zone

◆ Further away from the edge of the city there may be small villages. During the 1960s these became popular for people who preferred to live well away from the city and to travel in and out daily to their work. These are known as 'commuter settlements'.

◆ The main feature of these settlements is that their principal function is residential and very few people actually work in the settlement.

Putting all of these together, it is possible to produce various **models** of settlement which are an ideal representation of the structure of a settlement. Figure 2.2 shows the main models of settlement.

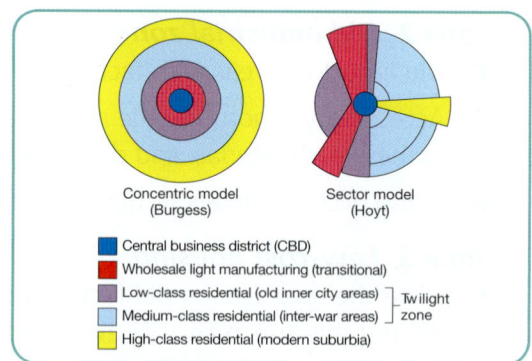

Concentric model (Burgess)   Sector model (Hoyt)

■ Central business district (CBD)
■ Wholesale light manufacturing (transitional)
■ Low-class residential (old inner city areas) ⎤ Twilight
■ Medium-class residential (inter-war areas) ⎦ zone
■ High-class residential (modern suburbia)

Figure 2.2 Models of urban settlement

◆ The first model is based on a series of rings from the centre to the outskirts. This is called the 'concentric model'.

◆ This model does not take into account the effect of local relief features such as hills and rivers.

◆ A second model called the 'sector model' does this.

◆ This model city allows for the fact that industries may develop alongside rivers or canals or major transport routes, and areas of low-cost housing may have been built beside the industrial areas.

- ◆ When drawn on a diagram these industrial and housing areas look like a wedge eating into the series of rings shown on the concentric model.

- ◆ It is likely that most settlements in Britain developed as a mixture of the 'concentric' and 'sector' models.

- ◆ Occasionally, questions may ask you about the city areas or zones and perhaps ask you to identify them on a map or draw comparisons between a particular settlement on a map and models which may be given to you on a diagram.

- ◆ Alternatively, you may be asked to compare different zones in a settlement from a given map.

- ◆ All of these areas with specific functions are called 'functional zones'.

Sample Questions and Answers

Question: General level 2002

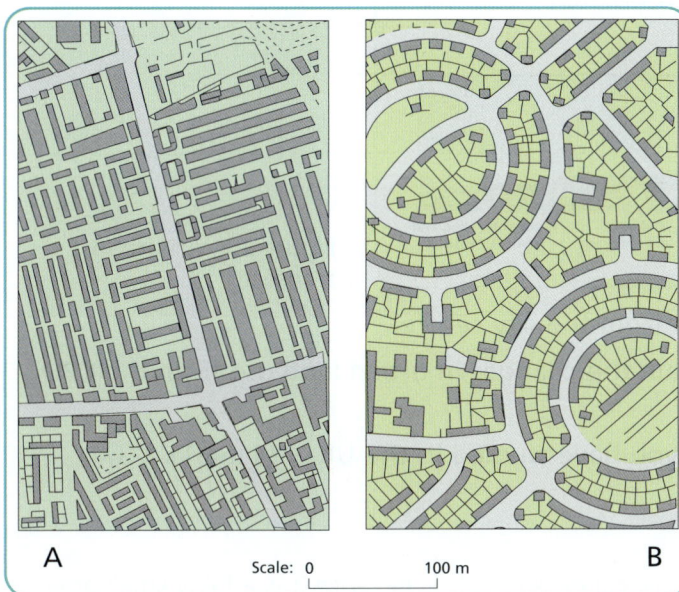

A

Scale: 0   100 m

B

Figure 2.3 Two housing areas in Liverpool

Study Figure 2.3.

Look at areas A and B.

Describe the differences in housing density and street patterns between the two areas. **(3 marks ES)**

Sample answer

Area A is much older and has a grid-iron (✔) street pattern whereas area B is newer and has a more curved, sometimes circular pattern (✔). In area B the houses are more spread out (✔) and there are fewer than in area A (✔).

69

Sample Questions and Answers continued

(?)

Comments and marks obtained

!

The answer correctly identifies differences in both street patterns and density. Two features of both are identified and the total marks obtained would have been 4 but there are only 3 marks available. The answer has more than enough to gain **3 marks out of 3**.

Question: Credit level 1997

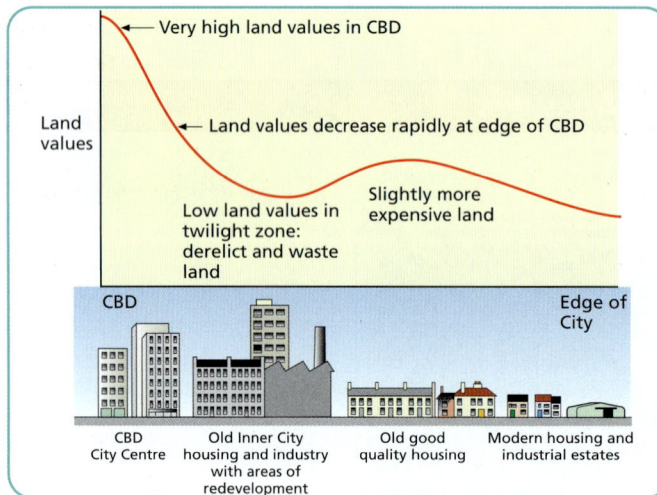

Figure 2.4 An urban transect

Cities have grown outwards from the city centre.

Look at Figure 2.4 and explain the changes in urban land use from the CBD to the edge of the city.

You may refer to examples you have studied in the UK. **(6 marks KU)**

Sample answer

In the CBD the land is extremely expensive. This is because it is a fairly small area where many businesses wish to locate (✔). It is in demand from major chain stores and so it is very expensive to buy land there (✔).

At the edge of the CBD the land is much cheaper as the major businesses have located in the CBD and the smaller ones wish to be on the outskirts and they cannot afford to pay high prices (✔). When the housing begins it is old tenements which have been redeveloped but are still very cheap and owned by local councils, e.g. old areas of Glasgow (✔).

These areas have cheap housing as people work in the city centre and need to have easy access to work (✔). After this it is better-quality housing and privately owned ones (✔). They are more modern and slightly more expensive (✔).

! **Comments and marks obtained**

This question asks for explanation but candidates often make the mistake of simply describing the different land uses from the centre of the settlement – the CBD – to the edge. Answers which do this instead of explaining would gain only a maximum of 2 marks out of a possible 6.

Although this answer does have some description, it does also make an effort to explain the differences in land use by referring to factors such as cost of land, the demand for land from businesses, cheapness of land further away from the centre, council houses being cheaper than privately owned houses, the need for workers to live near their work, and the modern, more expensive houses at the edge.

The answer even refers to a city in Britain – Glasgow. There is more than enough detail in this answer to merit the full **6 marks out of 6**.

Spheres of Influence

◆ The sphere of influence of a settlement is the area from which that settlement draws its trade or custom. Essentially it is the area served by that particular settlement.

◆ The greater the range and importance of the services contained in a settlement, the greater the area served by that settlement.

◆ A city with large department stores and perhaps regional council offices, will serve a larger area than a town which does not possess these functions.

◆ As the range of functions changes, for example, certain shops closing and new shopping areas opening, the sphere of influence may grow or contract.

◆ Not all spheres of influence are of the same size or shape. Apart from settlement size, other factors such as relief and accessibility can affect these features.

◆ Hilly and mountainous areas, for example, in the north and north-west of Scotland, can prevent people from visiting certain settlements because that involves longer journey times.

◆ Equally, the density of population directly affects the size and shape of spheres of influence.

◆ For example, for settlements in areas with a low population density such as highland areas, the sphere of influence may have a large physical size in terms of actual area served. In densely populated areas such as Central Scotland, the physical size of the sphere of influence will be much smaller but the population served will be greater.

◆ The sample question and answer below indicates the methods used to work out the size of a sphere of influence.

◆ The main basis for determining the extent of the sphere of influence includes factors such as:

– a survey of shoppers/consumers in a main shopping area using a questionnaire to find out where they live and how often they shop in that settlement

– an interview with the manager of a large furniture store to find out the delivery area of the store

– an interview with the manager of a local newspaper to find out the circulation area of the newspaper

– interviews with people in charge of main services such as hospital ambulance services, large secondary schools, garages, town planning offices.

◆ Once the information is gathered it can be plotted on a map to show the boundaries, or a composite boundary, of the service area/sphere of influence.

◆ Questions on this topic offer a good opportunity to test 'gathering' techniques related to spheres of influence at all three levels.

Sample Questions and Answers

Question: Credit level 2002

Describe the gathering techniques which could have been used to identify the sphere of influence of Inverness as shown in Figure 2.5.

Justify your choice of techniques.
(6 marks ES)

KEY

- - - Sphere of influence of Inverness

Upland area over 200 m

Towns and cities with over 10,000 people

- 500,000–1,000,000
- 100,000– 500,000
- 25,000– 100,000
- 10,000– 25,000

SHETLAND ISLANDS

ORKNEY ISLANDS

OUTER HEBRIDES

NORTH WEST HIGHLANDS

Elgin
Inverness
Aberdeen

GRAMPIAN MOUNTAINS

Dundee
Perth
Stirling
Glasgow
Edinburgh

N

0   100 km

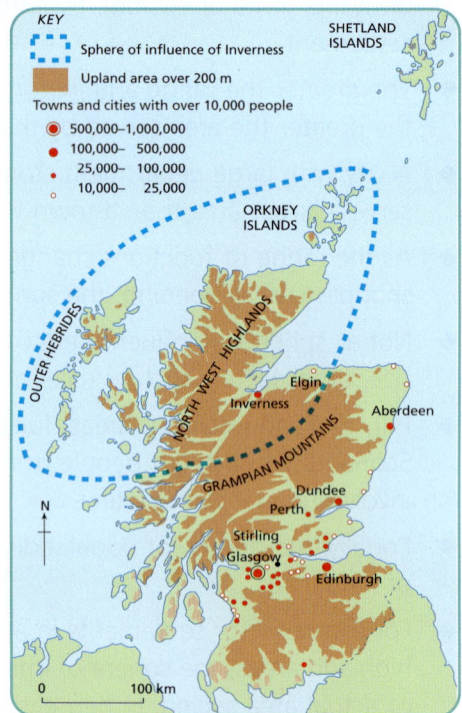

Figure 2.5 Sphere of influence of Inverness

Sample Questions and Answers continued ➤

Sample Questions and Answers continued

?

Sample answer

To find the sphere of influence of Inverness you could ask local shopkeepers where they deliver to (✓). This will provide accurate and reliable, up-to-date information (✓). Ask shoppers where they have travelled from to do their shopping (✓). Again this will provide accurate and reliable information. Look at car registrations to find out where they have come from. This provides good information about people and where they have been shopping.

Comments and marks obtained

The first two statements on delivery areas and interviewing, and the statement on reliable information, obtain 3 marks. However, the candidate has confused car registrations with car tax discs which would indicate where cars were registered and possibly where the shopper lives, so does not gain any further marks.

Unfortunately repeating statements on reliable information is not worth additional marks. The answer gains only **3 marks out of 5**.

Glossary of Associated Terms
Key Idea 7: Settlement

Central Business District (CBD): the zone of a town or city which contains the major shops, businesses, offices, restaurants, clubs and other entertainments and is normally located at the centre of the settlement, at the junction of the main roads.

Dereliction and decay: old buildings such as factories and houses which, through age and wear and tear, are no longer usable and have been abandoned. They cause visual pollution and are often demolished to make way for new developments.

Functional zones: areas of a settlement where certain functions are dominant, for example, industrial zones.

Functions: individual activities which settlements perform, such as commercial, industrial, administrative, transport, religious, medical, recreational and residential.

High order functions: the most important functions such as department stores, council offices and art galleries.

Low order functions: less important functions such as small corner shops, post offices and petrol stations.

Site: the actual land upon which a settlement was originally built.

Site factors: the factors which influenced people to choose a particular site, such as nearness to a water supply

73

(river), flat land for building, high land for defence, at a suitable point on a river where a bridge could be built, near raw materials.

Sphere of influence: the area from which a settlement draws its customers for various functions. Usually the size of the sphere of influence varies in direct response to the size of the settlement.

Urban areas: another term for towns and cities.

Urban model: an idealised view of the internal structure of a settlement. It can be used as a basis for comparing settlements.

Key Idea 8

Urban settlements have dynamic patterns relating to their size, form and function.

Key Point 19

Settlements, regardless of size, are continually changing. These changes may affect the structure, size (physical and population), function and individual zones within the settlement.

These changes often create problems which you should be able to identify.

You should also be able to discuss possible solutions to these problems, and their impact.

The main changes which have occurred recently in settlements include those affecting the following features of settlements.

Residential Areas

◆ The three main types of housing – low, medium and high-cost – have all changed significantly in most settlements in Britain.

◆ Many town and city councils have built areas of council houses both in the inner city and on the outskirts. These have included tenements, terraced houses and high-rise flats. Many residents have been given the opportunity to move from older houses

which have since been demolished to make way for new houses in council house schemes. This has often resulted in the break-up of communities and has had a serious impact on the social structure including the break-up of families.

◆ The styles of medium and high-cost houses have changed considerably. Many private housing estates have been built on the fringes of cities or in small villages on the outskirts, attracting people who are willing to commute to the city on a daily basis for work.

◆ More recently, city authorities have tried hard to attract people back to houses within the city and so help stop the problem of population decline.

◆ During the 1960s vast numbers of people moved from cities to new towns several kilometres away from the older urban areas. Not only did the new towns offer new housing but they also offered the prospect of new jobs since many modern industries were attracted to them.

◆ The councils of new towns often offered incentives for people to move there such as subsidies and rent-free accommodation to attract the industries. Many new towns such as East Kilbride, Crawley and Milton Keynes were very successful in attracting people and industry away from the older urban centres.

Central Business Districts

◆ There have been many important changes to this zone including the building of covered shopping centres and the closure of city centre shops and businesses due to a fall in the number of customers.

◆ There have also been improvements and changes to roads and streets to ease problems of traffic congestion.

◆ New office blocks have replaced older buildings such as bus and railway terminals.

◆ Cities which had port facilities, such as Glasgow, London and Liverpool, have had their former dockland areas radically altered.

◆ New conference centres, housing, business parks, museums and leisure complexes, such as Canary Wharf in London, have replaced old dockland and industrial areas.

Industrial Zones

◆ As older heavy industries declined during the 1970–90 period, these were often demolished leaving 'brownfield' sites within cities available for new developments such as housing, shopping centres and more modern industrial estates.

◆ Many new industries found sites on the outskirts of cities. These sites were often on cheap land with good access to nearby motorways.

Transport

◆ Most cities have seen major changes to their transport systems. New roads, motorways, by-passes and ring roads have been built around and through cities to cope with the very heavy increases in traffic.

◆ These changes have often resulted in the clearing of housing areas, industrial areas and the moving of people to other parts of the city.

◆ In many cases the environment has been improved. In some cases the new roads have led to increased congestion, danger and pollution levels.

◆ One-way systems, bus lanes and 'pedestrianised' zones have been tried in many cities in an attempt to reduce problems caused by traffic congestion. These efforts have often been very successful.

◆ Pedestrianised zones are areas which have been set aside purely for pedestrians. Normally traffic is not allowed to enter these areas, although delivery vans are often an exception to this. The idea is to make the area more pleasant for shopping, and to cut down on traffic congestion and pollution. Many cities in Britain now have these zones.

Sample Questions *and Answers*

Question: Credit level 2000

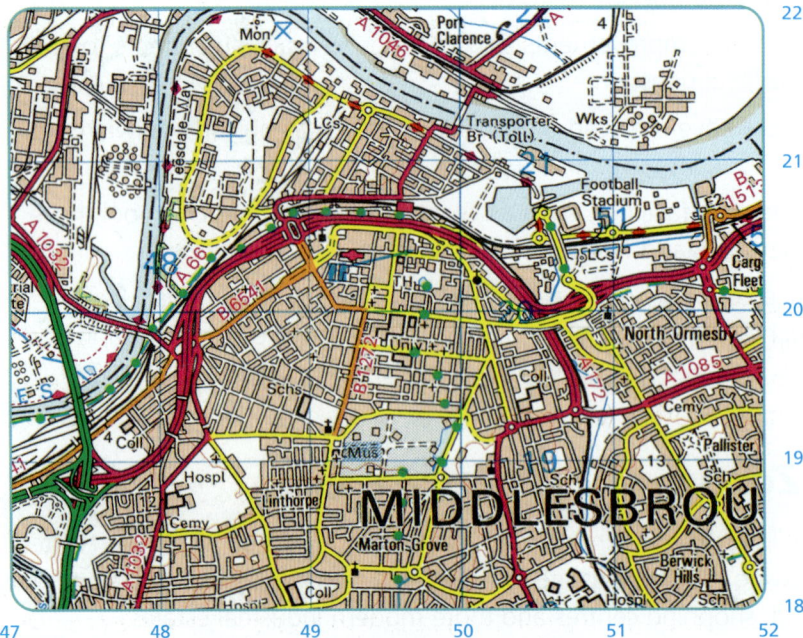

Figure 2.6a Ordnance Survey map of Middlesbrough

Sample Questions and Answers continued ?

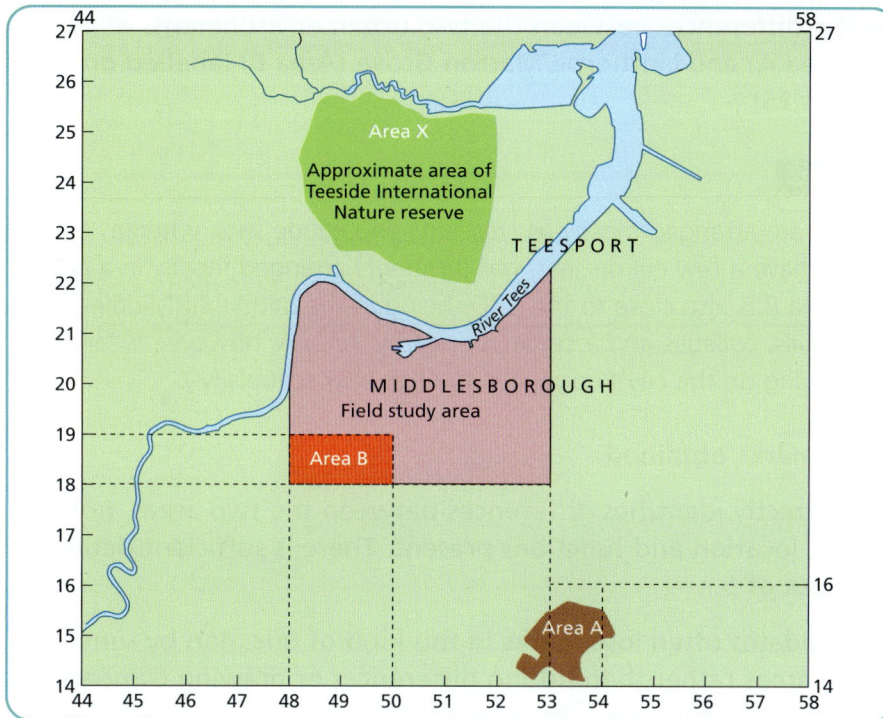

Figure 2.6b Middlesbrough: field study

This question refers to the Ordnance Survey map extract of the Middlesbrough area (Figure 2.6a) and Figure 2.6b.

a) Give the four-figure grid reference of the square which contains Middlesbrough's CBD (Central Business District).

Give reasons for your answer.   **(3 marks KU)**

Sample answer

Middlesbrough's CBD is in square 4920. (✓). Streets here and round about are mostly in a grid-iron pattern (✓), and the main road (A66) (✓) and other roads converge at this point. This square also contains the town hall, the railway and bus stations, an information office (✓✓) and 4 churches, with more in the surrounding squares.

Comments and marks obtained

The answer correctly identifies the grid square with the CBD and gives plenty of map evidence to support this, including references to street patterns, roads converging, and the presence of specified functions. The answer gains the full **3 marks out of 3.**

Sample Questions and Answers continued ?

b) Describe the differences between the two urban environments of Nunthorpe (Area A) and Linthorpe/Marton Grove (Area B) labelled on Figure 2.6b. **(5 marks ES)**

Sample answer

Area A's streets are arranged mostly as crescents and cul de sacs, whereas area B, though it does have a few cul de sacs, has its streets arranged mostly in a grid-iron pattern (✓). Area B is also close to the CBD, being in the middle of Middlesbrough (✓), with hospitals, colleges and a cemetery nearby. Area A, however, is far from the CBD (✓), being on the city's edge (✓), and has only schools (✓).

! Comments and marks obtained

This answer correctly identifies differences between the two areas, noting street patterns, location and functions present. There is sufficient detail to gain **5 marks out of 5**.

Note that candidates often lose marks in this kind of question by simply describing the areas rather than noting differences or drawing comparisons between them as the question asks them to do.

Glossary of Associated Terms
Key Idea 8: Settlement Change

Commuter: Those who live in commuter settlements.

Commuter settlement: small settlements on the outskirts of major towns and cities where people live but travel into the main settlement for employment and services.

Green belt: areas surrounding a city or town in which laws control development such as housing and industry in order to protect the countryside.

Inner city: the area near the centre which contains basically the CBD, the older manufacturing zone and zone of low-cost housing.

Overspill population: this refers to people who have moved out of the main city to smaller towns or new towns.

Pedestrianised zones: traffic-free areas within the city centre where people can shop and walk along streets where traffic is forbidden to enter.

Renewal and regeneration: the processes by which older areas are demolished and replaced by new buildings often having totally different functions to the original building or area.

Ring road: road built specifically to take traffic away from the city centre and to help solve the problem of congestion.

Suburb: housing zone on the outskirts of a town or city away from the busy central zone.

Traffic congestion: heavy build-up of traffic along major routes and within the city centre which causes great problems of cost and pollution for many cities in the UK.

Key Idea 9

Farming systems provide food and raw materials.

Key Point 20

This Key Idea refers to farming within the UK. Farms operate as systems with inputs, processes and outputs. You should be aware of the main characteristics of arable, pastoral and mixed farming types.

You should also know the factors which influence these farms, how they operate as a system and the main changes which have affected them over the past 50 years.

Many of the questions asked about this topic, at all three levels, are based on Ordnance Survey maps.

Farm Types

◆ **Arable farms** specialise in growing crops, mainly cereal crops in Britain, such as oats, barley, rapeseed and wheat.

◆ **Pastoral farms** specialise in the rearing of livestock such as sheep and cattle.

◆ **Mixed farming** is based on a combination of arable and pastoral although one or the other may dominate.

Regardless of the type of farm, decisions have to be taken as to how the farm should operate. The decisions on 'what to farm' are based on a combination of factors which can be classified as either physical or human factors.

Physical factors

◆ **Relief:** This is the height and shape of the land. Steep land is obviously not very useful for growing crops whereas flat or gently rolling land is much more suitable. Steep land is generally better used for raising livestock such as sheep. Landscapes which have a combination of steep and flat land are probably best used for mixed farming, with sheep, cattle or other livestock kept in one part of the farm, and the less steep land used to grow a variety of crops such as potatoes, grass and barley which is often used as fodder to feed the livestock.

◆ **Climate:** growing seasons for crops depend greatly on seasonal temperature changes. Rainfall amounts vary throughout the country with northern and western parts generally being wetter. Drier areas and warmer areas tend to be better suited to arable farming.

◆ **Soil fertility:** areas with rich, fertile soils such as wide river valleys with alluvial soil are more attractive for farmers who specialise in arable farming.

◆ **Geology:** depending on the underlying rock, some areas may be better drained than others and this will have an effect on what can be grown. Artificial drainage is often used in areas with clay soils allowing them to be developed for arable and mixed farming.

Human factors

◆ **Labour:** most farms used to depend on a good supply of labour, especially arable farms which required a workforce for a wide variety of purposes including ploughing, seeding, cultivating and harvesting.

◆ **Mechanisation** of farming has led to huge decreases in the labour force as workers have been replaced by machines such as combine harvesters and milking machines.

◆ **Markets:** farms with produce such as milk and other dairy products tend to be located close to their market, where these products can be sold on a frequent basis to the consumers. This is because the produce is highly perishable. Farms specialising in livestock for meat do not need to be situated so close to their markets.

◆ **Transport:** all farms require transport to take produce to market. More isolated, pastoral farms tend to need less transport since deliveries to market are less frequent.

◆ **Government influence:** national and international governments can have a major impact on the type of farming practised. Government policies and multinational agreements such as those set up within the European Union, on output, quotas, subsidies, grants, import and export restrictions, for example, can help or hinder farming in various ways.

◆ **Policies:** for example, on 'set-aside' land, means that large areas of farmland are left fallow or unused to prevent overproduction.

◆ **Agreements** such as the European Common Agricultural Policy (CAP) often meant that farmers grew certain crops instead of others and were offered guaranteed prices for their produce regardless of market prices.

◆ **Crops** grown depended on demand and perhaps surplus supplies of other crops.

In some questions you may be asked to look at a farm on an Ordnance Survey map extract and then give an opinion as to which type of farming might be best suited to the area.

Other questions may ask you about the advantages and disadvantages of having one type of farming as opposed to another.

To answer this kind of question, you should be thinking of the main factors which affect the choice of farming. Not all of these factors can be seen from the map, e.g. climate or government influence.

However, by looking at contour patterns, the situation of the area in relation to road, rail transport, availability of water, closeness to market and drainage patterns which indicate local geology (that is, the type of rocks present in the area), you should be able to make a reasonable case for one type of farming or another.

Remember: your choice might not be strictly accurate but you will gain marks for using appropriate map evidence to support your choice.

When referring to map evidence, give appropriate four-figure or six-figure grid references.

Farm Inputs and Outputs

Occasionally you will be asked questions about the type of farm based on data relating to inputs, outputs and processes, particularly at Foundation and General levels.

Table 2.1 gives an indication of the main inputs, processes and outputs from a farm system.

Table 2.1 The farm system

Inputs	Processes	Outputs
Climate, soil, relief, geology	Ploughing	Crops
Seeds	Sowing	
Animal feed		Livestock sales
Fertilisers	Fertilising	Young livestock for
Pesticides	Insecticide spraying	Livestock rearing
Dairy products		
Capital (land, buildings, equipment, machinery)	Livestock rearing	Livestock
Manure		
Labour	Transporting	Animal produce
Transport	Medical care	
	Breeding	
Marketing costs	Milking	Profits
Administrative costs	Harvesting	
Medical costs	Marketing	
	Maintenance	

Depending on the combination of these three elements, you should be able to decide on the type of farming and give reasons to support your choice.

For example, if the land is flat, there is a lot of machinery, there are good roads nearby and the distance to the main settlement is not too far, the main processes include ploughing, sowing and harvesting and the output is crops, then the type of farming is clearly arable.

Change in Farming and Farming Landscapes

As with other economic enterprises, farming has changed greatly over the last 50 years. These changes have affected methods, organisation, farm output, labour, farming landscapes and the overall status of farming within the economy.

Several factors have been important in bringing about these changes including:

◆ **Mechanisation:** increasing use of machinery such as tractors and combine harvesters has had a tremendous impact on farms. First, more machines meant less labour was needed. Field sizes had to increase to allow machines to operate. This resulted in hedges being removed and the size of farms increasing. Many farms amalgamated to increase in size. Buildings had to be built to store the machinery.

◆ **Improved technology:** farming methods have become increasingly more scientific with computers, advances in medical care for animals, new improved seeds, and chemical fertilisers and insecticides being used on a much wider scale. Farmers have been trained in agricultural colleges to use modern technology and methods to improve output, reduce costs and increase profits.

◆ **Diversification:** in order to reduce costs and increase the profitability of farms, farmers have increasingly added new non-farming land uses to their farms. Much of this is linked to the leisure and tourist industry. Areas of farms are now used as golf courses or bike/rally tracks. Cottages once occupied by farm workers, now made redundant, have been converted into holiday homes for tourists. Areas have been set aside for camping and caravan sites. Some farmhouses offer bed and breakfast accommodation. All of this makes the farmer less dependent on having his income solely from farming.

A typical question would provide graphs showing trends such as workforce, size of fields, output and number of farms. You might then be asked a question in two separate parts: first to **describe** the trends and second to **explain** these trends.

Other questions, both at General and Credit levels, may present a diagram of two farming landscapes at different periods in time, and ask you to either describe the changes or differences or explain the differences or changes which have occurred.

If you simply describe changes when asked to explain you will forfeit marks.

Although you can describe the main changes, if you are asked to explain you must give reasonable explanations as to why these changes or differences have occurred.

Usually, if you base your answer on increased mechanisation and diversification, and develop these points, you should gain most of the marks available.

Sample Questions and Answers   **?**

Question: General level 2000

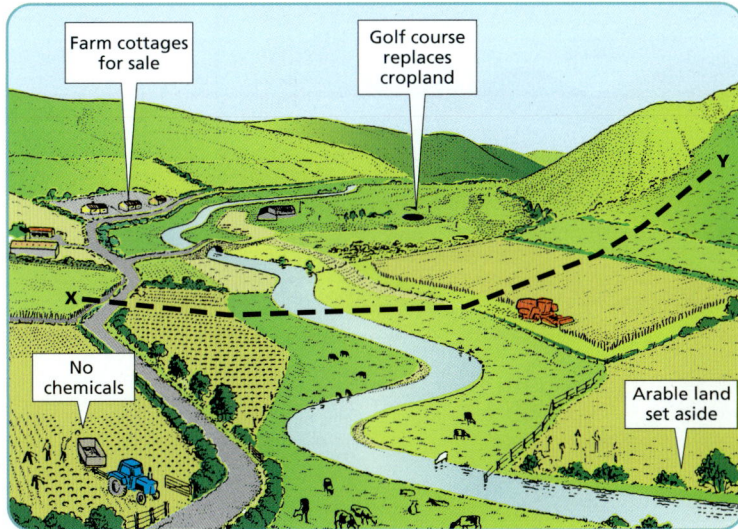

Farm cottages
for sale

Golf course
replaces
cropland

Y

X

No
chemicals

Arable land
set aside

Figure 2.7 Recent developments in agriculture

Choose two of the developments in agriculture shown in Figure 2.7 and give reasons why each one is taking place.   **(4 marks KU)**

Sample answer

Development: Farm cottages for sale.

Explanation: There used to be workers living in the cottages (✓) but the workers have been replaced by machinery (✓).

Development: Arable land set aside.

Explanation: The government doesn't want so many crops grown (✓) and gives the farmer money if he doesn't grow them and part of the farm is left untouched (✓).

! Comments and marks obtained

The first part of the answer gains 2 marks for referring to the former use of the cottages for workers and for saying that the workers have been replaced by machinery. The second part also gets 2 marks for referring to government intervention in not wanting certain crops to be grown and for giving the farmer money to leave parts of the farm untouched. The answer obtains **4 marks out of 4.**

Sample Questions and Answers continued

?

Question: Credit level 2003

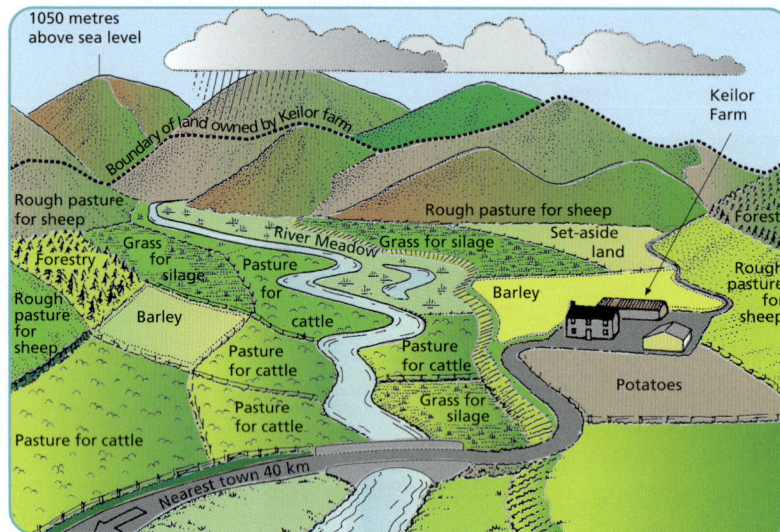

Figure 2.8 Sketch of Keilor Farm

Study Figure 2.8.

Keilor Farm is a hill farm producing mainly beef cattle and sheep.

Explain the links between land use and the physical and human factors affecting the farm. **(6 marks ES)**

Sample answer

The flat land surrounding the farm has been used for potatoes and barley which would need attendance regularly (✔). The hills furthest away are used for sheep which require no assistance; also this land is too steep for machinery (✔) so crops will not be grown. Down on the floodplain of the river it is mostly used for pasture for cattle (✔) as the farmer wouldn't want his crops ruined if it was to flood (✔). Also on the hills is forestry which provides shelter (✔). Any poor-quality land is used for grass to make silage (✔).

! Comments and marks obtained

Throughout this answer the candidate has tried to follow the basic instruction of the question, namely to explain the links between land use and the physical and human factors. The answer tends to concentrate on physical factors but does make reference to the use of machinery and nearness to the farmhouse, which are human factors. It refers correctly to flatness and steepness of land affecting land use and the possible flooding of the river and the use of the poorest land for growing grass. It could have referred to other human factors such as the distance to the nearest town and the reasons for the set-aside land.

Nevertheless the answer just manages to gain the full **6 marks out of 6**.

Glossary of Associated Terms
Key Idea 9: Farming Systems

Agribusiness: the operation of a large-scale farm which resembles a factory with large investments in farm property, maintenance, machinery and technology.

Arable farming: farms where the main activity and source of income is the growing of crops.

Casual workers: workers employed at specific times during the year, for example, at harvesting.

Cereal crops: grain crops such as oats, barley or wheat.

Common Agricultural Policy: a system used by the European Union to give farmers guaranteed prices for their products.

Crofting: a type of mixed farming found in northern Scotland which is not very profitable so farmers usually have to supplement their income by doing other part-time jobs.

Crop rotation: a system designed to maintain the fertility of soil by growing crops in different fields from time to time.

Dairying: farming in which the produce is derived from milk. Cows are reared to supply milk on a daily basis. This type of

farming has become highly mechanised with milking machines, high standards of hygiene and milk lorries taking the milk away to dairies to be processed into various products. Farmers are paid for the amount of milk supplied.

Diversification: adding different enterprises to the farm in order to improve income and allow the farmer to become less dependent on income from farm produce.

Drainage: if the underlying rock is clay, bog land and marshland may develop. Pipes are laid to drain excess water from the surface and allow the land to be farmed.

Farm system: the relationship between inputs, processes and outputs on a farm.

Fertiliser: substance which may be organic or chemical and is added to soil to increase fertility and improve crop yields.

Fodder crops: crops grown on a farm to feed animals, for example, grass and turnips.

Inputs: the basic needs of a farm before the farmer can begin to farm the land, such as seeds, livestock and machinery.

Insecticides: these may be sprayed onto crops to kill insects which may be attacking crops and therefore destroying the yield.

Mixed farm: farm which is based on a combination of arable and pastoral farming including, for example, dairy farms.

Outputs: the end product from the inputs and the processes of production on the farm.

Pastoral farm: farm which is based on rearing livestock such as beef cattle and upland sheep.

Pasture: land grazed by livestock. Some of this may be permanent, some temporary and some might be poor pasture which has been improved by, for example, underground drainage schemes.

Processes: the work done on a farm which obviously varies according to the type of farm.

Quotas: limits imposed on farmers in order to control the output of certain types of produce to avoid surpluses and therefore prevent a drop in prices.

Root crops: crops harvested for their roots, such as potatoes, turnips and carrots.

Rough grazing: poor-quality land used for grazing, for example, upland areas where soil is thin and is grazed by sheep.

Key Idea 10

The viability of manufacturing industry is affected by a variety of factors.

Key Point 21

You should be aware that industry can be classified into different groups and that each of these groups has similar factors which help determine the location of individual industries.

You should also make a point of knowing the main factors which influence the choice of location of industries.

Key Point 22

If you know these factors, you should be able to answer questions that ask about industry in an area of which you have no specific knowledge, for example, the Rhineland in Germany or north-east and south-east England.

Do not panic if this type of question appears in an examination.

Base your answer on your knowledge of location factors.

Many questions on industrial location are based on an Ordnance Survey map extract or diagrams which will contain a considerable amount of detail and information which you can use to support your opinions or statements.

Types of Industry

Industry simply means work which is done for economic gain. The various industries within the UK and Europe can be grouped into four main categories:

◆ **Primary:** these include 'extractive' industries such as fishing, forestry, farming, mining and quarrying. Essentially they take raw materials from the Earth and oceans.

◆ **Secondary:** these industries make or manufacture things from raw or semi-finished materials. These manufacturing industries include, for example, textiles, distilleries, furniture, electronics, petro-chemicals, oil refineries, car manufacturing, iron and steel industries, pharmaceuticals, and most items purchased in shops.

◆ **Tertiary industries:** these provide a service for the community in which they are located. They include, for example, local government services, transport, retailing and wholesaling, education and health, and trades such as hairdressers, plumbers and electricians.

◆ **Quaternary industries:** these are based on communications and rely heavily on information technology.

Factors Which Influence the Location of Industry

Industrial location factors can act either independently or in combination. In many instances certain factors may dominate.

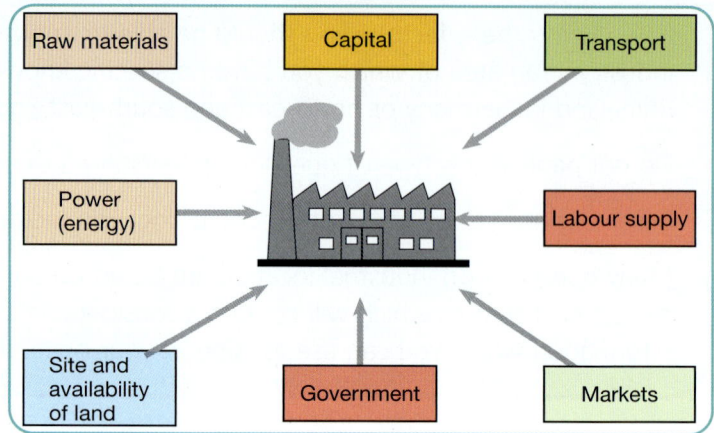

Figure 2.9 Industrial location factors

Raw materials

◆ In the past, about 150 years ago, when the major manufacturing industries of Britain and Europe were first developing, many of them had to be located near the source of raw materials such as iron ore, coal, limestone, stone, oil, farm produce.

◆ Such industries included iron and steel factories, coalmines, textiles, heavy engineering, food processing and oil refineries.

◆ This was usually because of the heavy additional costs involved in the transportation of the raw materials required.

Power (energy)

◆ Water and coal provided the main source of power for many older industries such as iron and steel and textiles.

◆ With the development of the national electricity grid this factor gradually became less important.

Labour supply

◆ Sometimes labour was attracted to the site of the industry, e.g. in the case of mining.

◆ In many cases industries had to be located where there was an adequate supply of workers, for example, in inner city areas.

Transport

◆ This usually takes the form of roads and railways.

◆ Without adequate transport facilities nearby, industries would find it impossible to operate due to the need to import raw materials, export finished products and bring workers into the industrial site.

◆ Transport has become one of most important factors influencing both old and modern industry, e.g. new industrial estates are often located near major motorway networks.

Markets

◆ Some industries locate as close to where they sell their products as possible, including, for example, retail and service industries.

◆ However, large multinational manufacturers often export their products either countrywide or even worldwide.

◆ Therefore, being close to their market is less important to them than for other industrial types.

Capital

◆ This is required to buy the actual site (land and buildings) on which the industry is located.

◆ It can be vitally important for all types of industry since many industries require large areas of flat land to build their buildings on, e.g. iron and steel factories, oil refineries.

◆ Many may require additional land for possible future expansion, e.g. a science park or retail park or modern industrial estate with a variety of light industries.

Government

◆ Should a government wish to attract industry to a particular part of the country where, for example, there is unemployment due to the decline or closure of older industries, they may offer financial incentives to prospective new industries.

◆ This can take the form of grants, subsidies, rent-free accommodation, retraining schemes for workers, low rates or taxes.

◆ This is a highly political factor and depends on the policies of the party in government. However, it can be the most important factor, especially in the case of new modern industrial estates often involving foreign companies.

◆ More recently countries within the EU have benefited from receiving grants and loans from the European regional funds for development areas, e.g. southern Italy, Ireland and Spain.

Personal decisions

◆ In some instances personal preferences by management or main investors may override other location factors and may help to explain the sites of some industries in unusual locations.

Industrial inertia

◆ This factor operates when the original location factors for a large industry, such as an iron and steel works, no longer apply.

◆ The industry may continue in its original site since to close it would involve too many problems such as redundancies, costs of moving machinery, and infrastructure.

Examination questions based on an Ordnance Survey map extract will ask you to explain the location of a particular industry using map evidence.

How to answer these questions

◆ If you can remember the main location factors you should inspect the map for suitable evidence to support your explanation.

◆ This may include reference to flat land (absence of contours), nearness to raw materials, accessibility due to the proximity of road and rail networks, being near a source of labour, e.g. a town or city, and room to expand and develop.

◆ Other questions may provide you with information on a map or diagram and ask you to evaluate the site of an industry in terms of suggesting the main advantages and disadvantages.

◆ In answering these questions make full use of the information given to support your answer.

◆ You will be given full credit for carefully selecting appropriate information from the given source.

◆ You could also either be asked questions on gathering data techniques on an industrial area or be given statistical information and asked about appropriate techniques for processing the data.

◆ The latter would involve referring to statistical diagrams such as pie charts, bar graphs, maps or tables.

◆ You will also be asked to give reasons (*justify*) why you think these techniques are appropriate.

◆ Your answer to a Credit level question should contain much more detail and may involve more complex techniques than in a General level question.

◆ At Foundation level you will be asked only to identify appropriate techniques, normally from a given list.

Sample Questions and Answers ?

Question: General level 1999

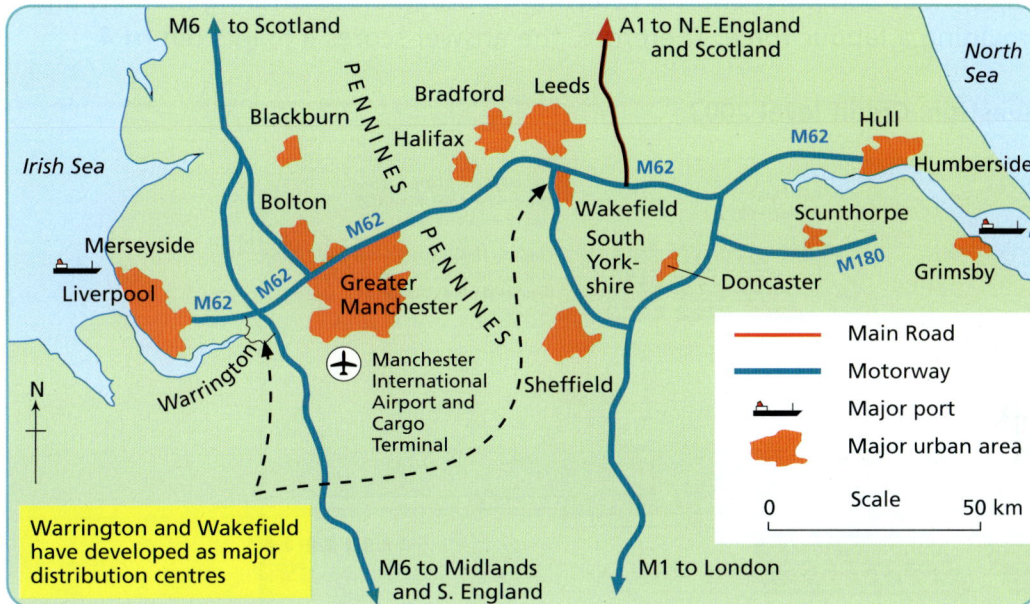

Figure 2.10 The M62 Corridor

Look at Figure 2.10.

The M62 motorway runs from Merseyside to Humberside. The area around it, known as the M62 corridor, is often described as England's new economic super region.

Give reasons to explain why so many companies are locating here.
(4 marks KU)

Sample answer

Because there is easy access to major roads like main roads, motorways (✔) and major ports, this is great for transport (✔).

They are also in major urban areas which is useful to sell products (✔) or find new workers (✔).

Comments and marks obtained

This is a good example of a question relating to an area which you may not have studied. That does not mean you cannot answer this question. All you need to know are the basic factors affecting industrial location and to use appropriate information from the resource diagram in the question. This

Sample Questions *and* Answers

answer does exactly that. It gains marks for referring to transport, noting major roads and ports and for referring to urban areas offering markets and providing a labour force. Therefore, the answer scores **4 marks out of 4**.

Question: Credit level 2003

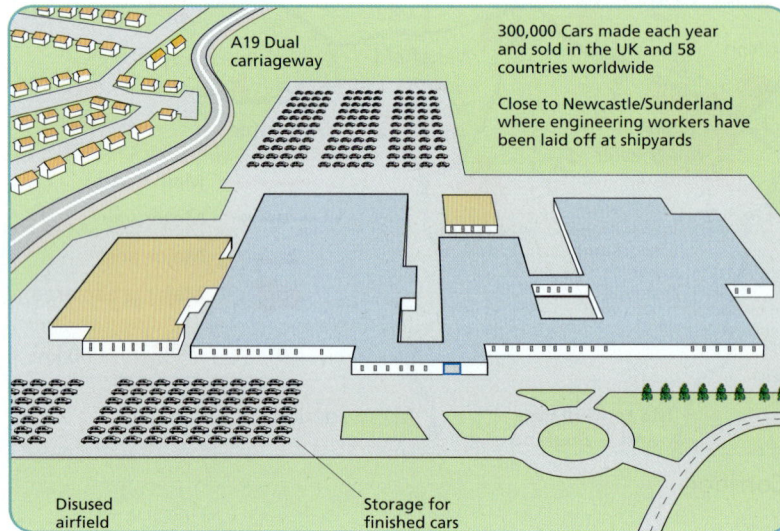

Figure 2.11 Nissan car factory, Washington (view looking south)

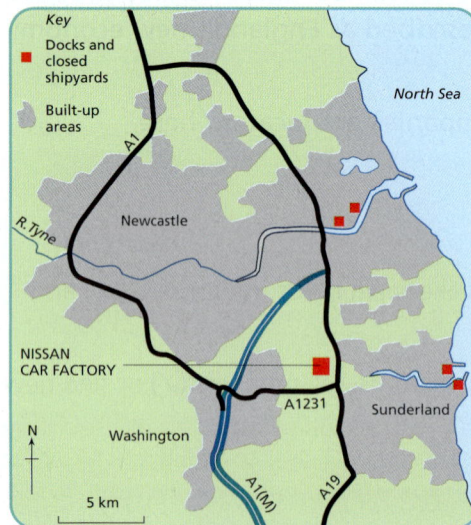

Figure 2.12 Location of Nissan car factory

Study Figures 2.11 and 2.12.

Suggest reasons why Nissan chose to locate their car manufacturing plant at this site in Washington, north-east England. **(6 marks ES)**

Sample Questions and Answers ❓

Sample answer

Nissan chose this site for their car factory because the factory is very close to the main roads (✔) which makes it easier to transport their cars around the country to different showrooms (✔).

The factory is also close to built-up areas which makes it easy for people to buy their cars.

There are also docks areas close to the factory which makes it easy to ship the cars to 58 countries worldwide (✔).

The area around the factory can be used for storing the cars. The land is not built on which means the car company might be able to expand the factory to bigger premises (✔).

The car plant will employ lots of people and the boatyard is closed so they could employ more people (✔).

Comments and marks obtained

This is quite a good answer which picks out relevant factors such as transport, market, site and labour. The first sentence gets 2 marks for developing the original point about transport. The vague reference to market (built-up areas) is not clear enough to obtain a mark. The point about storing cars and expansion is essentially the same point and therefore scores only 1 mark. The answer gains a total of **5 marks out of a possible 6**.

Glossary of Associated Terms
Key Idea 10: Industry

Business park: a new industrial estate usually occupied by offices or computer-based industries.

Enterprise zone: an area which receives government assistance to attract new industry and create new employment opportunities.

Extractive industry: primary industry which takes raw materials from the ground, e.g. mining, quarrying, fishing, forestry.

Greenfield site: land which has not been previously used for industry or any other buildings.

Heavy industry: industry which produces heavy-bulk materials, e.g. iron and steel, textiles, shipbuilding.

High-tech industry: industry which uses the latest technology to produce goods and services.

Industrial estate: an area set aside for modern, light industrial units, often located on the outskirts of towns/cities where land is available and the area is served by a good communication system, e.g. motorways.

Industrial inertia: this occurs when an industry remains in an area long after the original location factors no longer apply.

Light industry: industries which manufacture small, light-bulk products, e.g. window frames.

Manufacturing industry: known as secondary industries, which make a variety of products either finished or semi-finished.

Primary industry: industries which are based on extracting raw materials such as coal, ores, farm produce and forest products.

Service industry: also referred to as tertiary industries which provide services such as retailing, wholesaling, transport, legal, administrative and trades for communities.

Sunrise industries: new modern high-technology industries such as electronics. These industries are normally growing and gradually replacing older, declining industries.

Key Idea 11

Economic change has social and economic consequences.

Key Point 23

This Key Idea can be tested in various ways and in conjunction with other Key Ideas. For example, it is often examined within the topics of settlement, farming and industry.

Questions may involve looking either at an Ordnance Survey map or at diagrams which show industrial/settlement/agricultural areas in different time periods. You should be aware of the main types of economic change which can affect industry, farming and settlement.

Key Point 24

You may be asked to compare the old and more recently built industrial landscapes, to describe the changes which have occurred, or perhaps explain why the changes were necessary and what impact the changes may have had on the local community.

Economic Change

◆ The kind of changes which can affect **industry** may involve the demolition of old industrial buildings either because they are no longer needed or perhaps because they represent a danger or visual pollution of the environment. These buildings may have been replaced by newer more modern buildings and the whole area may have been completely redeveloped with new roads, better access, and a more pleasant looking environment.

◆ The impact of this change on the **local community** could, in the first instance, leave many workers unemployed due to the closure of the older industry. Many workers may have moved to different parts of the country in search of employment, while local shops and businesses could suffer from lack of demand through either people moving out of the area or having less money to spend due to being unemployed.

◆ **New industry** being brought into an area obviously stimulates the whole economy. It may involve retraining schemes for older workers, and new workers moving into the area bringing more custom for local businesses, including the housing and building industry.

◆ Often new industry is attracted through the efforts of local or national government which may offer **financial incentives** in the way of grants, subsidies and tax benefits.

◆ The community may also benefit from renewal of the local **infrastructure**, for example, improvements in local communications such as new motorways or bus and railway terminals.

Economic change in agriculture

◆ During the last 50 years there have been many changes in farming involving changes to farm size. Farms have become much bigger due to the removal of hedges and field sizes being increased.

◆ Much of this is due to mechanisation. This in turn has led to changes in the ways farms operate and the number of employees required.

◆ Often questions will focus on graphs showing trends in farming and will ask you to describe the pattern shown on the graphs and perhaps to explain why these changes have occurred.

◆ Another typical question may provide diagrams of two farming landscapes, the first from several decades ago and the second one which is much more up to date.

◆ You could be asked to explain the changes in the landscapes, for example, changing use of farm buildings such as farmworkers' cottages which have become holiday homes, new buildings built to house farm machinery, some land now used for non-farming purposes such as camping and caravan sites or mountain biking, which is related to the process of diversification.

Economic change in cities and towns

◆ There are many new industrial locations within and outside towns. The trend is towards 'out-of-town' shopping centres, leisure and recreation sports centres.

◆ You may be asked to comment on the impact of these changes on the environment, for example, additional land which was used for farming is now used for industrial or recreational purposes, leading perhaps to increased traffic congestion and possibly increased pollution levels.

◆ Most questions examining this Key Idea are resourced based, that is on a map, diagram, graphs or statistics. In other words they examine Enquiry Skills.

◆ You will not necessarily be tested on your actual knowledge but rather on your skills in using the resources provided to form an opinion, make choices or suggest possible consequences of the changes referred to in the resource.

◆ At General level questions on this topic are usually worth 4 marks and at Credit level they could be worth 5 or 6 marks.

◆ Clearly the difference between the two levels is the amount of detail which you can provide in your answer.

Sample Questions and Answers

Question: General level 1997

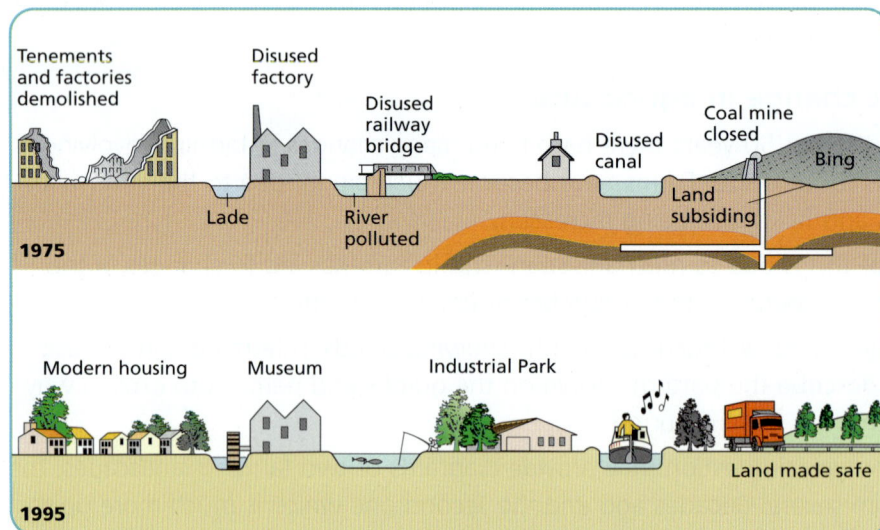

Figure 2.13 Industrial land use, 1975 and 1995

Look at Figure 2.13.

Describe the ways in which the industrial landscape has changed between 1975 and 1995.   **(3 marks ES)**

Sample Questions and Answers continued

Sample answer

Modern housing has been built to replace demolished tenements (✓) and a disused factory has been converted into a museum (✓). The polluted river has been cleared and made suitable for fishing (✓). The disused canal has been put into use for barges (✓).

Comments and marks obtained

The reason why this is such a good answer deserving full marks is that the answer does not simply state a change, but is very specific on what changes have occurred, e.g. modern housing/demolished tenements, disused factory/museum, polluted river/suitable for fishing and finally disused canal/use for barges.

The candidate has made very good use of the resource provided and gains **3 marks out of 3**.

Question: Credit level 2000

In the 1950s, Eire was a country based on agriculture and characterised by emigration and a fairly low standard of living.

Figure 2.14

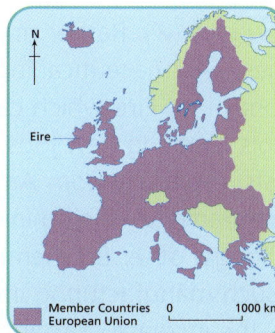

Figure 2.15 Location of Eire in the EU

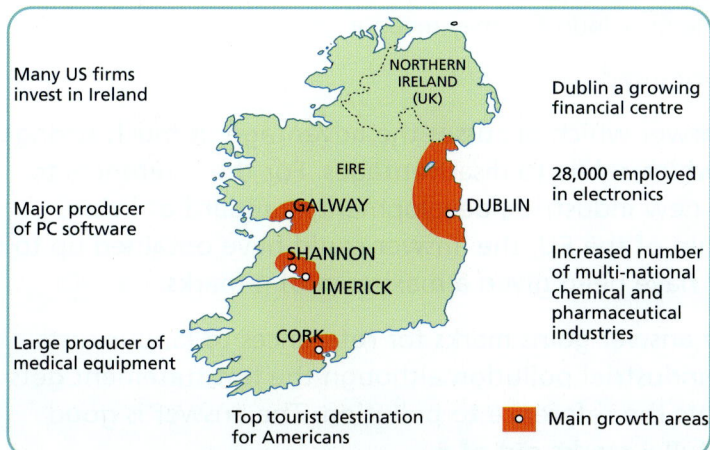

Many US firms invest in Ireland

Major producer of PC software

Large producer of medical equipment

Top tourist destination for Americans

NORTHERN IRELAND (UK)

EIRE

GALWAY

SHANNON

LIMERICK

CORK

DUBLIN

Dublin a growing financial centre

28,000 employed in electronics

Increased number of multi-national chemical and pharmaceutical industries

Main growth areas

Figure 2.16 Eire in the late 1990s

Sample Questions and **Answers** continued **?**

- ◆ Labour costs low by EU Standards
- ◆ Many university graduates
- ◆ Favourable tax system
- ◆ High quality environment

Figure 2.17 Why firms are going to Eire

Study Figures 2.14–2.17.

'Over the last thirty years the economy of Eire has changed as many foreign firms have chosen to locate there.'

Describe in detail the advantages and disadvantages which economic change has brought to Eire.   **(6 marks ES)**

Sample answer

The change of Eire's economy bringing many foreign firms into the country has created many new jobs (✔). Rather than depend on agriculture, Eire now has chemical industries, pharmaceutical industries and computer industries (✔). This brings lots of money into Eire which can allow it to develop further (✔), improving homes and schools, creating new jobs, raising the standard of living (✔). The increase of tourists, especially from America, will also bring money and jobs, e.g. in hotels and restaurants (✔). Being in the EU, Eire can trade its produce easily with no taxes amongst other EU members (✔). However, with labour costs being low, workers in such manufacturing industries do not receive a great deal of money (✔). Also, the increase of tourists may damage Eire's scenery through litter and erosion (✔), and new industries also bring pollution with them (✔). Being a lot of industry, especially chemical industry, may mean a lot of pollution.

Comments and marks obtained

The first part of the answer which discusses the advantages is much stronger than the second part which refers to disadvantages. For the references to new jobs, examples of new industries, development, standard of living, tourism and membership of the EU, the answer could have obtained up to 6 marks, but would only have been given a maximum of 4 marks.

The second part of the answer gains marks for references to labour costs, damaged scenery and industrial pollution although the last statement gets no marks since it repeats the reference to pollution. The answer is good enough to obtain the full **6 marks out of 6**.

Glossary of Associated Terms
Key Idea 11: Economic Change

Business park: industrial estates with businesses which may sell products or provide services directly to the public.

Dereliction and decay: this refers to abandoned buildings, perhaps mines, offices or industries, which have resulted from closures, and often they become a source of visual pollution if they are not demolished.

Diversification: the process whereby an economic enterprise such as a farm takes on a range of additional activities to increase profits, for example, renting land for golf courses, caravan sites, paintball enterprises, quad biking.

Economic: relating to financial developments.

Economic effects: the financial impact of change on, for example, employment, incomes, running costs, building costs and costs to the local community.

Environmental consequences: the effects of change on physical and human environments, for example, changing land use, improvements and bad effects on the environment such as increased pollution, or changes to the population caused by people moving to or from areas as a result of changes such as industrial closures or new industry being built.

Industrial decline: this happens when, for various reasons, industries in an area have to close. It has largely involved large traditional, heavy industries such as mining, iron and steel making, textiles and shipbuilding.

Industrial growth: this happens when an industry increases its output, the number of workers and its profits and usually has a positive effect on the local community.

Multiplier effect: the wider effect of change on other activities such as other industries, settlements or rural activities. For example, when a major industry closes this may cause local shops and other businesses to close since people made unemployed have less money to spend and may move elsewhere to seek new work.

Science parks: industrial areas which are closely connected to technological institutions and universities and are often involved in producing high-technology products. They often help to replace declining industries and offer new job opportunities to workers from closed industries following retraining schemes.

INTERNATIONAL ISSUES

Key Idea 12

Population is unevenly distributed.

Key Point 25

There are many parts of the world which, for a variety of reasons, attract large numbers of people. There are also other areas which are much less attractive.

You should be aware of the factors which either attract or repel population. If you know these factors in detail you will be able to attempt any question which asks you to explain population distribution in any given area.

Key Point 26

Since this Key Idea is tested within a global context, the questions can be set in any area of the world, for example, Bolivia, Peru, Pakistan, Japan or the UK.

It is not necessary for you to know about these countries in detail. If asked to explain population distribution you will be given plenty of information in maps or diagrams on which you can base your answer.

Population Distribution and Density

Positive factors

Factors attracting large numbers of people to an area include:

◆ flat land for building settlements
◆ a good supply of water
◆ availability of raw materials for developing industry
◆ a good climate to live and work in
◆ an area which is easily accessed, e.g. in a valley or on a coastal plain
◆ good fertile soil for farming.

Negative factors

Factors which tend to cause population to avoid areas include:

◆ land which is too high or steep to build on, e.g. mountainous areas
◆ areas where the climate is inhospitable, e.g. too hot, too cold, too wet or too dry, such as hot deserts, tundra, ice caps and tropical rainforests
◆ areas which lack resources
◆ areas which are fairly inaccessible, e.g. mountains, jungles, deserts
◆ areas which are remote.

Explaining population distribution

When asked to explain patterns of distribution you should:

◆ describe where the main areas of population occur or perhaps where the density of population is greatest or least
◆ use the data given to you to help explain possible attractive features or any obvious negative features of the area, referring to some or all of the factors listed above
◆ avoid simply providing a list of factors
◆ try to link the factors together, e.g. type of relief and climate, making the area suitable for settlement, industry and farming.

Sample Questions and Answers (?)

Question: General level 1997

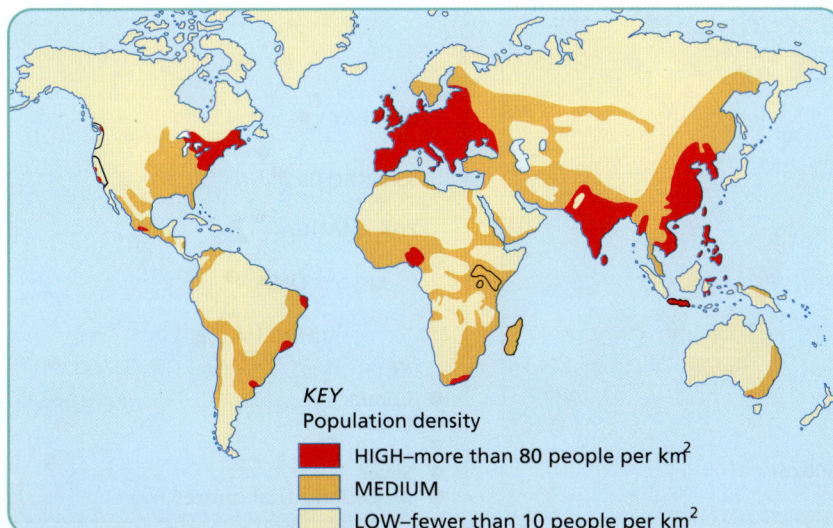

KEY
Population density

■ HIGH–more than 80 people per km^2
■ MEDIUM
□ LOW–fewer than 10 people per km^2

Figure 3.1 World population distribution

Sample Questions and Answers continued

Look at Figure 3.1.

Explain why some parts of the world are more densely populated than others.

Refer to both physical and human factors in your answer. **(4 marks KU)**

Sample answer

Some parts of the world are more densely populated than others because some areas have plenty of resources (✓) and good flat land for building on (✓). Also some countries are richer than others, attracting more people to them. Good climates also attract people (✓) and that is why some areas are more densely populated than others:

Comments and marks obtained

This is a good answer but will gain only **3 marks out of the 4** available. The points which gain the marks are indicated (✓). However, the reference to richer countries does not merit a further mark since many poor areas and countries in the world have a high population density, e.g. India and South East Asia.

Question: Credit level 2000

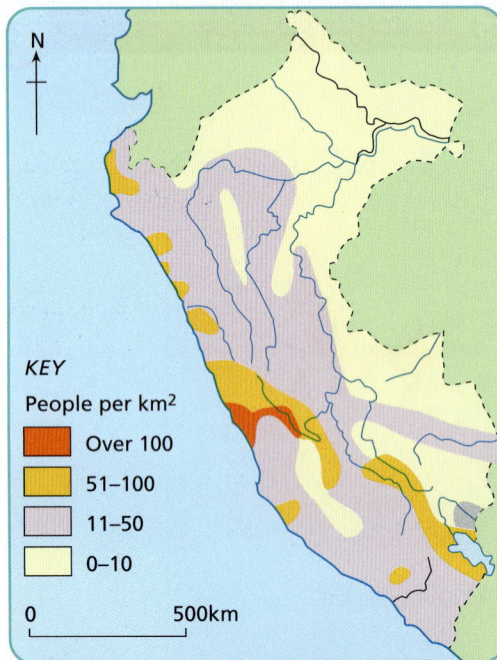

KEY

People per km²

- Over 100
- 51–100
- 11–50
- 0–10

0   500km

Figure 3.2 Population density of Peru

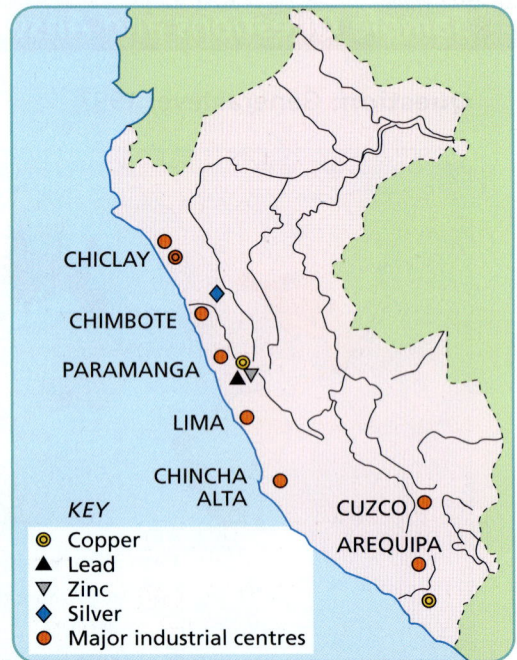

CHICLAY

CHIMBOTE

PARAMANGA

LIMA

CHINCHA ALTA

CUZCO

AREQUIPA

KEY

- ◎ Copper
- ▲ Lead
- ▽ Zinc
- ◆ Silver
- ● Major industrial centres

Figure 3.3 Resources and industry of Peru

Sample Questions and Answers continued

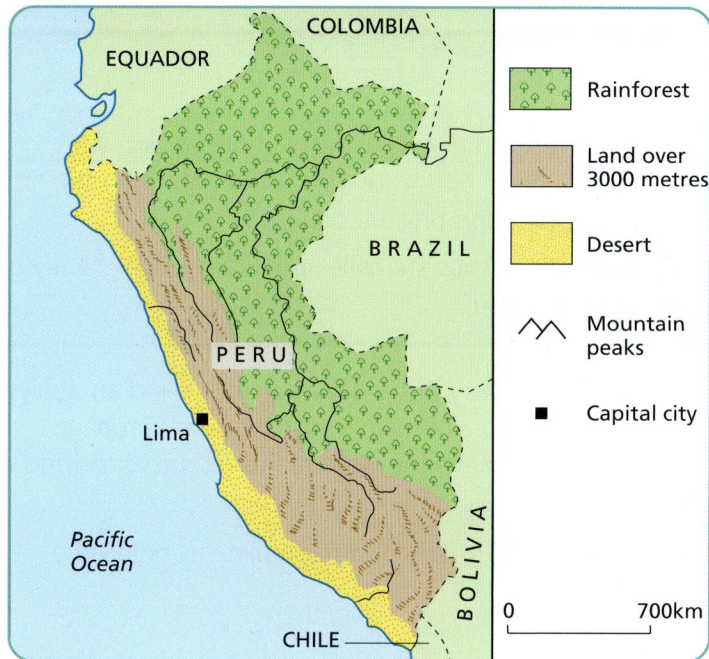

Figure 3.4 Physical features of Peru

Look at Figures 3.2–3.4.

Explain the population distribution of Peru. **(4 marks KU)**

Sample answer

The population distribution of Peru does not seem to rely too much on physical factors though the avoided areas are mostly rainforests (✓) and mountain peaks where living conditions are harsh (✓). The most densely populated area is around the capital Lima where many will stay and work (✓). The other most densely populated areas (with 51–100 people per km^2) are around resources (✓) and industries, since work is provided in these areas (✓), e.g. copper and industrial sites in the north and the rich area around Lima containing copper, lead and zinc (✓).

Comments and marks obtained

This is an excellent answer which contains a number of relevant points. The candidate makes good use of the information provided in the question and refers to both physical and human factors. The answer also gives good examples which help to develop the points made, e.g. the reference to mineral resources. The answer easily gains the full **4 marks out of 4**.

Key Idea 13

Populations have measurable social and economic characteristics.

Key Point 27

◆ There are various terms which refer to different aspects of population with which you should be familiar including: birth rate, death rate, growth rate, life expectancy, infant mortality rate, demographic model.

◆ The data used in calculating these statistics are collected in a national survey of the population called the national census.

Not every country in the world is able to carry out a detailed census and an analysis of its population. Developed countries such as Britain, most Western European countries, the USA, Canada and Australia have the wealth and resources to fund a full national census once every decade.

Poorer, less developed countries often find it difficult to complete as detailed a census due to problems including:

◆ low literacy rates
◆ inaccessible areas
◆ sections of the population who are nomadic – that is, not settled for any length of time in any given place
◆ lack of money to pay for census workers and the technology required
◆ wars and natural disasters.

As well as being aware of the different ways of measuring population characteristics you should be able to explain the factors affecting these characteristics. You should also be able to suggest reasons why populations can differ so greatly from one country to another.

Factors Affecting Population Characteristics

◆ In less developed countries, birth rates and death rates are greatly affected by social, economic, traditional and religious factors. In general, higher birth and death rates occur in the least developed countries of the world.

◆ Social factors such as poor health care, poor education systems, poor food supplies and lack of a balanced diet, and poor infrastructures – especially in the availability of clean water supplies and efficient sewerage systems – often lead to high death rates.

◆ Large families are also common for reasons such as religion and tradition, e.g. avoiding the use of birth control methods, the need for children to provide a cheap labour supply and insurance for parents in old age.

◆ Lack of money in some countries prevents countries from investing in education, health, farming, industry and good housing which would lead to a healthier population.

◆ High infant mortality rates are common in less developed countries due to a lack of health provision and high birth rates.

◆ In more developed countries low birth and death rates are common.

◆ People tend to marry later in life and postpone having children for economic reasons, for example, the wife wishes to continue working and may have fewer children.

◆ Birth control methods and advice are readily available and are used extensively.

◆ High standards of health care and of living in general allow people to live longer so lowering the death rate and increasing life expectancy rates.

◆ For similar reasons infant mortality rates are lower in more developed countries such as the UK and the USA.

The Demographic Transition Model

The demographic transition model is a diagram which summarises the main changes in birth rates, death rates and overall population over time – Figure 3.5.

This model consists of four separate stages and line graphs which indicate changes in birth rates, death rates and growth rates over different periods of time.

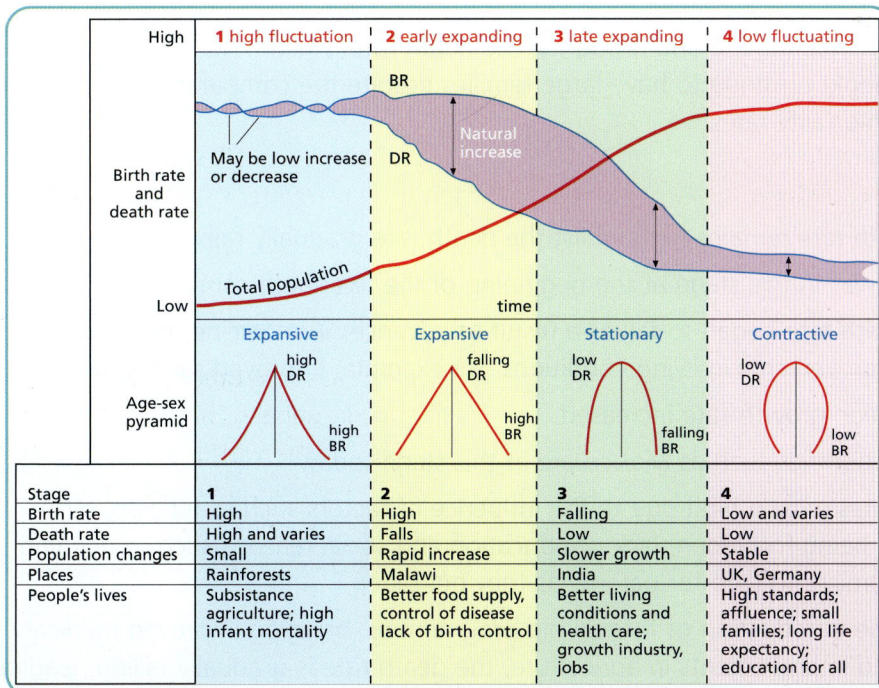

Stage	1	2	3	4
Birth rate	High	High	Falling	Low and varies
Death rate	High and varies	Falls	Low	Low
Population changes	Small	Rapid increase	Slower growth	Stable
Places	Rainforests	Malawi	India	UK, Germany
People's lives	Subsistance agriculture; high infant mortality	Better food supply, control of disease lack of birth control	Better living conditions and health care; growth industry, jobs	High standards; affluence; small families; long life expectancy; education for all

Figure 3.5 The demographic transition model

The demographic transition model summarises patterns of population growth in various countries throughout the world and is based on the relationship between births and deaths. You should be able to explain what is happening in each of the four stages of the model and give reasons why this is occurring.

Stage 1

◆ Both birth and death rates are high and therefore population growth is low.

◆ Britain was in this stage about 250 years ago at about the start of the period known as the Industrial Revolution.

◆ At this time industries were developing in towns and cities and attracted people from rural areas.

◆ The larger the family the higher the income from children working in factories. Living conditions were poor with the majority of people living in overcrowded tenements with poor sanitation. Many died from diseases which today would be regarded as fairly minor.

◆ At present there are several countries throughout the world where birth and death rates are both high including, for example, less developed countries such as Ethiopia.

◆ Through war, famine and droughts, the death rate is high and life expectancy is low. People continue to have large families despite the comparatively high infant mortality rates.

Stage 2

◆ The birth rate remains high whilst the death rate gradually falls.

◆ Britain was in this stage at the beginning of the twentieth century.

◆ The death rate began to fall as a result of advances in medicine, better diets and improvements in living conditions, for example, improved sanitation.

◆ Population growth rate increased.

◆ Some countries such as Mexico are at this stage today.

◆ Birth rates remain high due to the influence of factors such as religion, birth control being forbidden, lack of education and social reasons such as needing large families to care for parents as they get older.

◆ With the introduction of foreign aid programmes, bringing improved medical care and improvements in agriculture, the death rate is gradually falling, leading to rapid increase in population growth.

Stage 3

◆ The death rate continues to fall and the birth rate also begins to fall but at a much slower rate.

◆ Population continues to grow.

◆ Through long-term aid projects provided through the United Nations agencies, many less developed countries are beginning to reduce both their birth and death rates.

◆ This includes countries in Asia, Africa and South America.

◆ Medical assistance is provided in the fight against various diseases such as cholera, typhoid and malaria, and diseases caused by hunger and malnutrition.

Stage 4

◆ Death rates continue to fall in more developed countries of the world, with advances in medical science and new drugs becoming available to treat major diseases such as cancer and heart disease.

◆ Birth rates also continue to fall due to changes in lifestyle, with artificial contraception being widely available and widely used.

◆ With many women wishing to continue to work rather than having a large family as in previous generations, birth rates and death rates are almost equal.

◆ As a result population growth is very small.

◆ Countries in this stage may find that their population consists mainly of people in the older age groups.

You will rarely be asked detailed questions on this topic in the Foundation examination.

However, questions on birth and death rates, their causes and effects, and questions related to the demographic transition model, are fairly common at both General and Credit levels.

Most of these questions are based on your knowledge of these topics. You may also be asked to express opinions and discuss the advantages and disadvantages of population growth based on appropriate resources such as diagrams or tables.

Your answer should develop the points given in the resource. Good answers avoid simply taking information directly from diagrams without providing additional comment.

Sample Questions and Answers

Question: Credit level 2003

Figure 3.6 Location of Bolivia

Look at Figures 3.6 and 3.7.

a) Give reasons why it may be difficult to take an accurate census in a developing country such as Bolivia. **(5 marks KU)**

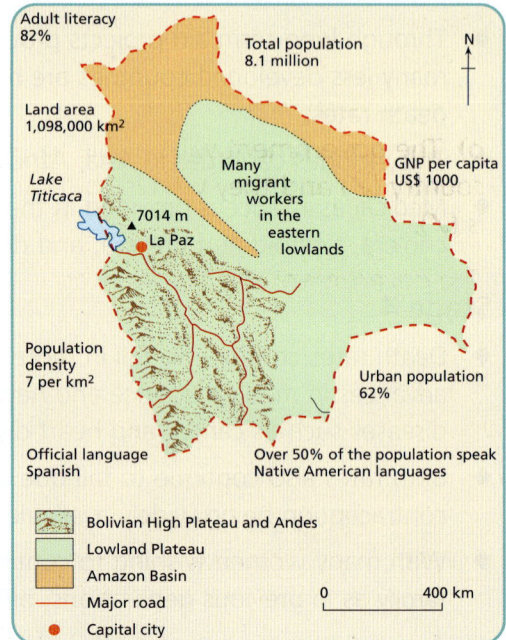

Figure 3.7 Bolivia

Sample answer

a) The population density is only 7%/km² so people are very spread out and difficult to get in touch with (✓). Half the population speaks Native American and only 82% of the adults are literate (✓), so some will not be able to fill in a census form as they can't read or write (✓). 62% of the people live in an urban area while other people live in the eastern lowlands or in upland areas which causes problems when carrying out a census (✓). There are also no main roads in the Amazon Basin or Lowland Plateau so some people will be unable to get a census form to fill out (✓). There are 8.1 million people altogether and to find all these people would be very difficult.

Comments and marks obtained

The answer contains many good points such as the population is spread out, half of the population speak a Native American language, only 82% being literate and many unable to complete census forms. The references to people in the eastern lowlands and lack of roads and therefore being unable to receive a form are also worth additional marks. The answer therefore obtains a total of **5 marks out of 5**.

Sample Questions and **Answers** continued ➤

Sample Questions and Answers continued ?

b) What use could the government of a developing country make of population census data? (4 marks KU)

Sample answer

b) The government will be able to understand how many people are living in the country (✓) and they will also be able to understand where most of the population is (✓).

The government will be able to decide where to build schemes which would help the people in that area (✓) and the government could build houses which are cheap enough for poor people to live in (✓).

Comments and marks obtained

Although the answer is not well written, nevertheless there are four accurate statements which gain a mark each. References to the number of people, where they are and building schemes and cheap houses to help the population all gain marks. This gives a total of **4 marks out of 4**.

Key Idea 14

In any area the size and structure of the population are subject to change.

Key Point 28

These changes are due to variations in birth and death rates and the effects of in-migration and out-migration.

Population structures are shown by diagrams called population pyramids which consist of bar graphs showing the age and sex distribution of the population.

Key Point 29

From inspection of these graphs you should be able to detect patterns of high and low birth rates, high and low death rates, and high and low life expectancy rates.

You should also be able to classify countries as developed (more economically developed countries, or MEDCs) or developing (less economically developed countries, or LEDCs) according to their population structures.

Figure 3.8 in the sample question opposite shows structures which are typical of MEDCs and LEDCs.

◆ The first pyramid has a fairly narrow base with more people in the older age groups in both the male and female sections. The shape has a bulge in the middle area and is relatively wide in the older age groups.

◆ This indicates a low birth rate, a low death rate and high life expectancy.

◆ In effect the graph shows an ageing population.

◆ Unless the birth rate increases or more migrants with young families come into the country, this country may face several problems in years to come.

◆ With fewer children this may have an impact on the future workforce.

◆ With more older people in the population, this will increase demands for more health care, pensions, homes that older people can retire to and pressure on younger people, for example, with increased taxes.

◆ The second pyramid has a wider base indicating a high birth rate.

◆ The shape begins to get narrower until it is very narrow at the top in the 65+ age groups.

◆ This shows that life expectancy is low.

◆ This is typical of an LEDC.

◆ With high birth rates, there will be many children to feed and look after.

◆ Countries with this structure are often poor and therefore cannot afford good health care for their population.

◆ There is greater demand for education but lack of money to make it widely available.

◆ Due to poverty and low standards of living, including housing and food supply, the average life expectancy is low.

Sample Questions and Answers

Question: Credit level 1997

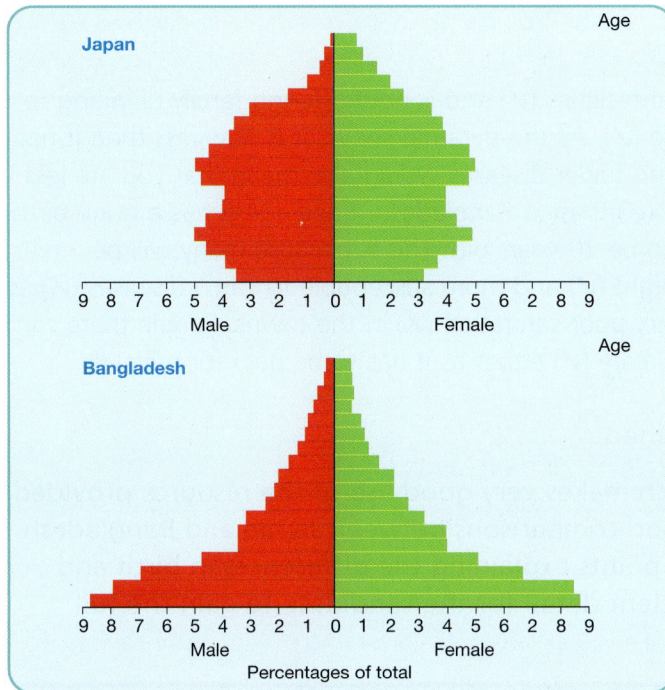

Figure 3.8 Population pyramids for Japan and Bangladesh

Study Figure 3.8.

a) Describe the differences between the population structures of Japan and Bangladesh. **(3 marks ES)**

Sample answer

a) Japan has few people in the very young age groups whereas Bangladesh has a lot of people in the age groups 0–10 (✔). There are more people in the 20–60 age groups in Japan than in Bangladesh (✔). Bangladesh has fewer people living beyond 70 than Japan (✔). This means that people live longer in Japan than in Bangladesh, that is life expectancy is higher (✔).

Comments and marks obtained

The answer correctly identifies three main differences in the population structures by referring to the lower, middle and upper parts of each pyramid. It gains further marks for the references to longer life and higher life expectancy. The answer is worth the full **3 marks out of 3**.

Sample Questions and Answers continued ➤

Sample Questions and **Answers** continued

b) Give reasons for the differences shown in (a).   **(6 marks KU)**

Sample answer

b) Japan has better medical facilities (✓) and is now offering family planning to try and reduce the birth rate (✓). As the pyramid shows it is now less than it has been (✓). Better nutrition and fewer diseases would also mean that you are less likely to die young in Japan (✓) than in Bangladesh. Bangladesh has a huge birth rate but less than half live to be 20 years old. This is because many will be malnourished and underweight (✓) and more susceptible to catch diseases (✓) as the country probably has very poor sanitation (✓) in the towns. People there can probably not afford medical care (✓) either as it has to be paid for privately.

Comments and marks obtained

This is a good answer which makes very good use of the resource provided. The answer also makes good comparisons between Japan and Bangladesh and makes many relevant points explaining the differences in birth and death rates. There is sufficient detail for the candidate to gain the full **6 marks out of 6.**

Migration and Migration Patterns

◆ **Migration** is the movement of people from country to country or from continent to continent.

◆ **Emigration** occurs when people leave a country to live in another country.

◆ **Immigration** happens when people enter a country from another country to live there.

Reasons for migration

There are a number of reasons why people may wish to migrate either from one part of a country to another part, or from one country or continent to another.

◆ In some cases people may not have a choice in the matter. They may have been forced to move, as in the case of the African slave trade of the seventeenth and eighteenth centuries.

◆ Similar movements have occurred during periods of war, for example, the First and Second World Wars during which large masses of people were displaced from their homelands to other countries.

◆ For the most part people migrate from one area to another in the hope that they might improve their standard of living and general lifestyle.

◆ They may move from an area of high unemployment, low wages, poor housing conditions, poor health and educational facilities, political or religious persecution.

◆ Hopefully the area to which they are moving may provide much better conditions than the area which they have left.

◆ The conditions which tend to create migration are called **push factors** and those which attract people to new areas are termed **pull factors**.

◆ When people move from countryside to urban areas (towns and cities) these factors are called **rural push** and **urban pull**. These are shown in Figure 3.9.

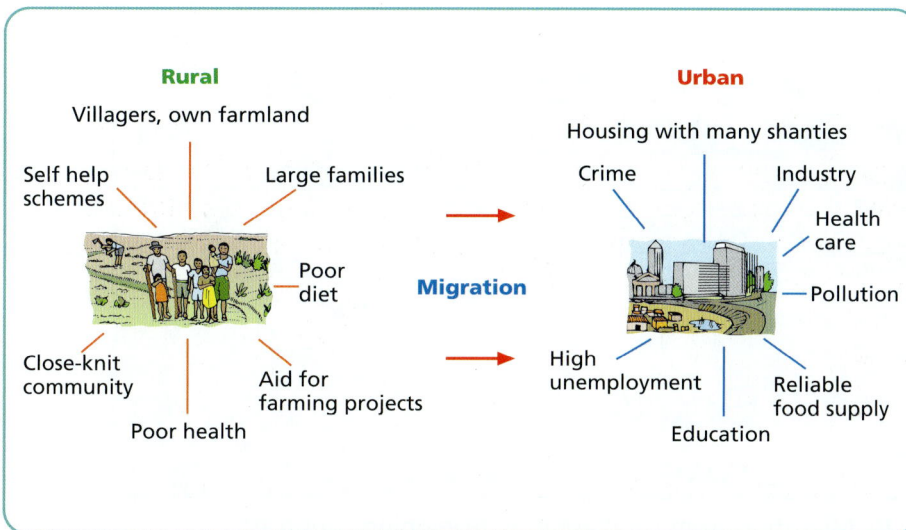

Figure 3.9 The developing world migration model

◆ This kind of pattern of movement is often seen in less developed countries which have a large portion of their population both living and working in rural/agricultural areas, for example, in India, Brazil and South East Asia.

◆ Often when immigrants move into an area they are disappointed and disillusioned when they realise that their expectations have not been fulfilled and many finish up living in worse conditions than those which they left.

◆ Many can only find accommodation in shanty towns on the outskirts of major cities in LEDCs.

◆ These places often have the poorest of living conditions, including poor housing, no amenities such as clean water supplies or electricity, and only the most basic sanitation.

◆ These conditions often lead to diseases such as cholera and typhoid which can kill the weakest in the population: the old and the very young.

Many examination questions are based on the reasons for migration, particularly from rural to urban areas, and also the problems faced by people who migrate.

Unfortunately, many candidates lose marks for not answering the question. For example, when answering questions on push and pull factors they instead spend a lot of the time answering the question on 'why they may be disappointed by their move to another country or move from rural to urban areas'.

Sample Questions and Answers ?

Question: General level 2003

Figure 3.10 Migration from rural areas in developing countries

Look at Figure 3.10.

Do you think people benefit by moving from the countryside to the city?

Explain your answer.   **(4 marks ES)**

Sample answer

Yes, because people get better jobs which lead to better pay (✓). There are good services like schools (✓) so therefore better education (✓), better standard of living, e.g. flats (✓) and better health care which leads to longer life (✓).

! Comments and marks obtained

Answers to this kind of question could refer to yes and no situations since the diagram contains data on both. This answer has concentrated on 'yes' points. It refers to employment and pay, better services and education,

Sample Questions and Answers *continued* ?

improved standard of living with an example and finally better health care leading to greater life expectancy. All these valid and well developed points give a total of **4 marks out of a possible 4**.

Question: Credit level 1999

'In developing countries people leave the countryside for a variety of reasons. They are attracted by the city and pushed from the countryside. Most find life hard in the city but large numbers continue to move there.'

Figure 3.11

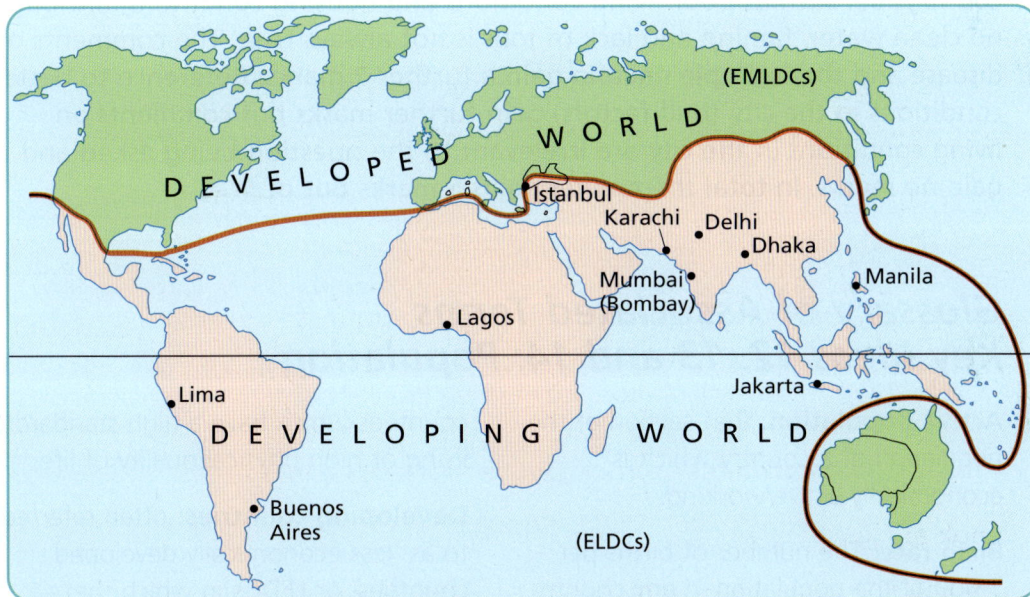

Figure 3.12 The ten fastest growing cities

Study Figures 3.11–3.13.

Despite the problems outlined in Figure 3.13, large numbers of people in developing countries are still moving from the countryside to the cities.

Explain why this is happening.
(6 marks KU)

Higher living costs

Not enough formal jobs

Shortage of good housing

Pollution and overcrowding

Figure 3.13 Problems of fast-growing cities in the developing countries

Sample Questions and Answers continued

Sample answer

People are still moving because of the poor living conditions in the countryside (✓) such as no clean water, famine and lack of jobs. Housing conditions are often poor (✓) and there is a lot of disease (✓) such as malaria (✓). People think that they will get a better-paid job in the city (✓) but when they get there they often live in much poorer conditions. They often have to live in shanty towns where there is a lot of disease and poverty.

Comments and marks obtained

The answer begins well by referring to the reasons why people leave the countryside, such as poor living conditions and housing conditions, although no clean water, famine and lack of jobs is not always true. The comments on disease and the example malaria gain a further 2 marks. Reference to better conditions in the city (pull factors) gain further marks but comments on living conditions in the city are irrelevant to the question being asked and gain no marks. In total the answer gains **5 marks out of 6**.

Glossary of Associated Terms
Key Ideas 12, 13 and 14: Population

Active population: that section of the population of a country which is economically active/working.

Birth rate: The number of births per 1,000 of the population in any country in any year.

Census: a numerical count of the population which is financed and carried out by the government at set periods of time, e.g. 10-year intervals.

Death rate: The number of deaths per 1,000 of the population of any country in any given year.

Developed country: sometimes referred to as 'more economically developed countries' or MEDCs; includes countries which have a high standard of living or high physical quality of life.

Developing countries: often referred to as 'less economically developed countries' or LEDCs in which the population generally has a low standard of living.

Empty lands: areas of the world which have a low population density, e.g. mountains and deserts.

Environmental factors: factors such as climate, relief, soil and water supply which can influence the distribution of population in an area.

Gross Domestic Product (GDP): the value of all goods and services of a country produced in one year. This measure is used as an indicator of the wealth of a country. However, it does not always reveal how well spread the wealth is among the population in general.

Growth rate: the difference between the birth and the death rate. It can be expressed as a percentage of the population. If births exceed deaths, clearly the population is growing. If the opposite happens the population is decreasing.

Infant mortality: the number of deaths of infants (below the age of 1 year) expressed as a rate per 1,000 of the population.

Life expectancy: the average age a person can expect to live to in any given country. This is a good indicator of level of development, since people in more

developed countries tend to live longer due to better health care, better diets, higher standards of education, housing and living.

Population density: the average number of people within a given area, e.g. 100 per square kilometre/km^2.

Population growth model: this shows different stages of population growth based on the relationship between birth and death rates.

Population structure: the grouping of the population of a country by age and sex. Inspection of the structure may also indicate trends in birth and death rates, life expectancy and the possible impact of factors such as war and migration on the population.

Standard of living: the level of economic well-being of people in a country.

Key Idea 15

International relations are dominated by a limited number of countries acting in conjunction with others.

Key Point 30

Countries throughout the world may be described as either developed/more economically developed countries (MEDCs) or developing/less economically developed countries (LEDCs).

The level of development of countries can be determined by referring to a set of economic or social indicators.

Usually it is very difficult to define the level of development of a country on one single indicator. It is better to base your judgment on a combination of indicators.

Key Point 31

You should be aware of the main economic and social indicators used to define level of development and be able to use these to classify countries in terms of being 'economically more developed' or 'economically less developed'.

Economic Indicators

◆ **Gross National Product (GNP):** the percentage of the working population employed in agriculture or industry, average per capita income, consumption of electricity per capita (kilowatts per capita), percentage unemployment.

Social Indicators

◆ Reference to population characteristics such as birth rates, death rates, life expectancy rates, infant mortality rates, crude growth rates, population structure, literacy rate, percentage of population with access to clean water, percentage of homeless, percentage of population attending primary/secondary education, average calorie intake per person.

◆ These indicators can be combined to produce a **Physical Quality of Life Index** (PQLI).

Questions on this topic may ask you to identify the level of development of certain given countries on the basis of selected indicators, to explain differences in the level of development or consequences arising from lack of development, or to describe methods of improving standards of development.

Key Point 32

You should know the main types of international alliance and be able to discuss the advantages and disadvantages to member countries.

International Alliances

◆ Those countries which have a wide range of resources (such as minerals, water, climate, industrial and agricultural output), combined with a highly developed level of technology (the USA, Japan, and the main countries of Western Europe such as the UK, France and Germany), tend to dominate world affairs. They have great influence economically and politically throughout the world.

◆ Often these countries act together as political or trading partners in alliances such as the North American Trade Alliance or the European Union.

◆ As a result they are in a position to dictate favourable terms to other countries, particularly those which are less developed.

◆ Often international companies from the dominant countries have exploited the resources of less developed countries and have used these to increase wealth and prosperity in the richer parts of the world.

Many questions, especially at General and Credit levels, have concentrated on the advantages of countries being a member of an alliance such as the European Union.

Advantages of membership

◆ Having a wider market in which countries can trade.

◆ Trading with other countries on more favourable terms.

◆ Avoiding paying tariffs and import taxes.

◆ Being able to trade as a large group with countries outside the alliance.

◆ Being able to obtain grants and subsidies from the alliance to help poorer areas within member countries to develop.

◆ The possibility of developing political and defence links between member states.

◆ Freedom of movement of labour by not having to apply for a permit to work in member countries.

◆ A common currency such as the euro means currencies do not have to constantly change from one country to another.

Disadvantages of membership

◆ Loss of historical identity through, for example, giving up passports.

◆ Severing historical links with trading partners throughout the world, for example, Britain and New Zealand.

◆ Having to submit to the general laws and regulations of the alliance.

◆ Being unable to trade on favourable terms on an individual basis.

◆ Competition for jobs due to freedom of movement of labour.

◆ Having to pay higher taxes to provide funds to pay for the administration and various grants issued by the alliance.

◆ Having to subsidise economically weaker members.

◆ Possibly increased prices due to agreements between member states.

◆ Loss of political power.

◆ Possibly subsidising inefficient industry and agriculture in other countries.

◆ Loss of own individual currency.

Sample Questions *and* Answers

Question: General level 2002

Figure 3.14 Government considering entry to a trade alliance

Look at Figure 3.14.

What are the advantages and disadvantages for countries which are members of a trading alliance? **(4 marks ES)**

Sample answer

An advantage is that the countries have a wider market to sell their products (✔). But this may cause competition between countries who produce the same thing (✔). There might be tension between countries or fear that foreigners may take their jobs (✔).

Comments and marks obtained

The answer does not make full use of the resource provided. It ignores good points such as aid to poorer regions, currency exchange rates, and disadvantages such as being told what to do and loss of passports. There are sufficient points on wider market, increased competition and fear of loss of jobs to obtain a total of **3 marks out of a possible 4**.

Sample Questions and Answers continued

Question: Credit level 2000

Table 3.1 Selected data for five countries

	Land area (km²)	Population (million)	GDP ($/head)	Exports ($ million)	Imports ($ million)	Urban (%)
USA	9,100,000	255	22,470	428	499	75
Brazil	8,400,000	150	2300	31	21	74
Japan	374,000	124	19,000	314	236	77
India	2,900,000	882	380	20	25	25
Russia	17,000,000	149	2240	58	43	74

Look at Table 3.1.

'Population size and land area are the main indicators which show the international importance of a country.'

Do you agree with the above statement? Give reasons for your answer.
(4 marks ES)

Sample answer

No, population size and land area do not show the international importance of a country. Take India for example: a large area (2,900,000 km²), a huge population (882 million) yet only $380 per head for GDP (✓), $20 million for exports and $25 million for imports (✓), the lowest in the table (✓). The real indicators of international importance are wealth (✓): GDP and exports and imports (✓).

Comments and marks obtained

This is a high-quality answer, although the introductory statement is not worth any marks. Marks are gained for noting that despite its size India has the lowest import and export figures, with examples. Further marks are gained for arguing that the real indicators are wealth, again with good examples. The answer is therefore worth the full **4 marks out of 4**.

The statistics given in Table 3.1 provide a good opportunity to ask a question on processing techniques. This frequently happens when data like this is provided as a resource.

Sample Questions and Answers continued

Question: Credit level 2000

Look at Table 3.1.

Suggest two methods you could use to show relationships between different data in the table.

Justify your choices in detail.   **(6 marks ES)**

Sample answer

The relationship between exports and imports can be shown on a scatter graph (✓). This shows the general trend between two sets of figures (✓) which in this case should be positive (as one increases so does the other) (✓).

A pie chart can be used to show the population in urban areas as well as the population in rural areas (✓). Taking the percentage of urban dwellers away from 100% gives the percentage of rural areas. (✓) This can be shown on a pie chart, which lets you see at a glance the fraction of the population (✓) in both, and just how much they are when compared to each other (✓).

Comments and marks obtained

The answer correctly identifies two appropriate techniques for 2 marks. The references to showing trends and positive/negative relationships gain a further 2 marks. The final comments on the qualities of the pie chart – working out the percentages and seeing at a glance the fraction of the urban/rural populations – and the comparison between them gains further marks.

This is an excellent answer, which shows a clear understanding of the appropriate use of the identified techniques for the data in the table and it gains the full **6 out of 6 marks**.

Glossary of Associated Terms
Key Idea 15: International Relations

Alliances: countries grouping together for various reasons. They include four main types: defence, trade, selling and social alliances.

Defence alliances: countries grouping together to form alliances to defend and protect each member, e.g.

North Atlantic Treaty Organisation (NATO).

Selling alliances: countries grouping together to get and maintain the best prices for certain products such as oil, e.g. Oil Producing and Exporting Countries (OPEC).

Social alliances: groups of countries with historical links such as the Commonwealth countries (former members of the British Empire).

Trade alliances: formed to benefit member countries in trade agreements, e.g. European Union.

Key Idea 16

Regions of the world are linked through trade.

Key Point 33

Trade is the exchange of goods and services between one country and another country or group of countries. Links between the more economically developed countries (MEDCs) and less economically developed countries (LEDCs) are established through trade. You must be able to look at trade figures and describe the pattern of trade between countries.

Describing Patterns of Trade Between Countries

◆ Patterns of trade are best described by referring to the different types of exports and imports which comprise the trade between countries or regions of the world. Types of exports and imports include raw materials, finished and semi-finished products. These are known as 'visible products'.

◆ Another form of trade is payments for services such as insurance payments, interest on loans, transport services or legal services provided by one country for another country. These are referred to as 'invisible' payments.

◆ This becomes important when thinking of 'balance of trade situations'. Balance of trade is the difference between exports and imports.

◆ Countries do well economically if they export more than they import. They do not do so well if they import more than they export because they are basically paying out more than they are earning.

◆ When invisible figures are added this can often turn a negative balance into a very healthy balance. Britain benefits greatly from this.

◆ Trade throughout the world is not always fair and equal. Many of the questions asked at General and Credit levels examine this idea and may ask for an opinion based on data provided in the question.

◆ Effectively, the poorer countries of the world have been providing the wealthier countries with resources such as oil, timber, agricultural and mining resources, often at low prices.

◆ These materials are used in manufacturing industries and made into finished or semi-finished products which are sold on at a much higher price, often back to the poorer countries. A good example of this is coffee.

HOW TO PASS STANDARD GRADE GEOGRAPHY

- Many LEDCs become dependent on the MEDCs, often ending up in debt to them.

- It is not easy for the poorer countries to change the situation, for example, by building factories to manufacture their own products.

- The LEDCs may be locked into trading agreements with multinational companies which favour the richer countries and to some extent exploit the developing countries.

- The developing countries cannot afford to withdraw from these agreements since they would lose the vital income which they receive from exporting their raw material resources.

- Close inspection of trade figures not only reveals the pattern of trade but also gives a clue as to whether the trade pattern is fair and equal.

Sample Questions and Answers

Question: Credit level 2000

Figure 3.15 Comparison of traditional and Fair Trade coffee prices

Perfect Coffee and How to Make It

- Warm the pot, add one heaped spoon of freshly roasted Fair Trade coffee.
- Filter out the middleman.
- Provide coffee farmers with a regular income.
- Help turn mud and straw houses into bricks and mortar.
- Enjoy your coffee . . . and help build a brighter future for the next generation.
- A better deal guaranteed for coffee growers.

Figure 3.16 Extract from a Fair Trade advertisement

Sample Questions and Answers continued

Study Figures 3.15 and 3.16.

Explain how the idea of Fair Trade is a way of helping coffee growers in the developing world. **(4 marks ES)**

Sample answer

In fair trade the profit made on coffee does not go to the shop which sells it (✓) (the middleman) but to the coffee grower himself (✓). The coffee may be dearer but at least the coffee grower gets a lot more than his usual 2p a jar with the rest going to the person who merely sells the coffee (✓). Coffee growers get a regular income for their produce (✓), with a guaranteed minimum price should sales go badly (✓), and this helps give the coffee growers a better standard of living (✓), e.g. improving their homes.

Comments and marks obtained

The answer covers several good points and actually scores more marks than are available. Good references are made to profit not going to shops but to the grower and an increase on the profit, guaranteed prices and better standards of living. The answer clearly obtains **4 marks out of 4**.

Question: Credit level 2002

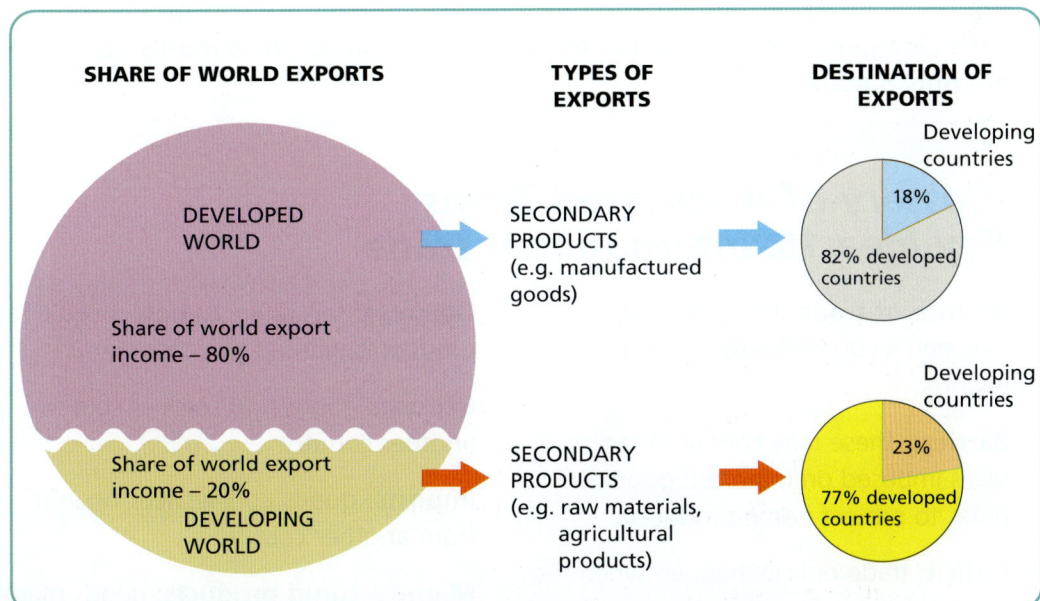

Figure 3.17 World trade

125

Sample Questions and Answers continued ?

Study Figure 3.17.

'The pattern of world trade benefits only countries of the developed world.'

Do you agree with this statement?

Give reasons for your answer.   **(5 marks ES)**

Sample answer

Yes I agree with the statement because in the diagram we are told that the developed country receives 80% income from its goods (✓) but the developing country only receives 20% for its goods (✓). Also, out of all the goods the developed world trades, only 18% goes to the developing countries (✓) and the rest goes to the developed countries (✓). Also from the trade given by the developing countries they only receive 23% but the developed countries get 77% (✓).

Overall developed countries get more trade but the developing countries need it most (✓).

Comments and marks obtained

Although the answer is not very well written, nevertheless the candidate is able to extract the appropriate data from the information to support his/her opinion. The answer contrasts imports and exports and the imbalance in the figures so far as developed and developing countries are concerned and in each case gains a mark. Overall the answer actually earns **6 marks but there are only 5 marks available**.

Glossary of Associated Terms
Key Idea 16: International Trade

Balance of trade: the difference between a country's exports and imports.

Barriers: these may consist of tariffs or taxes imposed on imported goods in order to protect home producers.

Deficit: trade deficits happen when the cost of imports is greater than money earned from exports.

Demand: this refers to how much of a product is wanted by consumers.

Exports: goods and services sold to another country.

Imports: goods and services bought from another country.

Manufactured products: goods made from raw or semi-finished materials, e.g. cars, textiles, machinery.

Sample Questions and Answers continued

Multinational companies: large companies with branches operating in countries throughout the world, e.g. BP, Shell, Unilever, Coca-Cola.

Overproduction: this happens when countries produce more than they can sell and as a result the price of the goods falls.

Producers: companies or countries which manufacture or supply goods or raw materials, e.g. Kenya is a producer of tea, BP produces petroleum and other oil-based products.

Surpluses: trade surpluses occur when the value of exports exceeds that of imports. Countries with a trade surplus normally have a sound economy and increasing wealth.

Key Idea 17

Schemes of self-help, along with national and international aid, seek to encourage social and economic development.

Key Point 34

You should know the different types of foreign aid schemes which exist throughout the world such as bilateral aid, multilateral aid, voluntary aid, tied aid, short-term aid and long-term aid schemes.

Types of Aid

Aid between countries takes a variety of forms. Some may be in the form of money.

Aid may also take the form of food and other supplies such as machinery and various types of medical supplies. Some might consist of medical assistance such as doctors and other medical staff.

Other aid may consist of the work of people trained in various fields such as agriculture, industry and education. Table 3.2 indicates the various types of aid on offer throughout the world.

Questions at General and Credit levels often ask you to identify types of aid from given information and to explain how you arrived at your identification.

Other questions may ask you to explain or describe the advantages and disadvantages of certain types of aid schemes as shown in Table 3.2.

A frequently asked question is one which asks you to decide which type of aid would be best suited to specific situations which are given to you in resources accompanying the question.

Table 3.2 Types of aid

Types of aid	Examples
Short-term aid	Aid which is required immediately, for example, following a natural disaster such as a flood or earthquake
Long-term aid	Aid which is given to countries over a long period of time to help them develop their economy or infrastructure.
Voluntary aid	Aid which is usually given through charities such as Oxfam, Sciaf and the Red Cross.
Multilateral aid	Aid from international organisations such as the United Nations supplied through development programmes.
Bilateral aid	Aid given from one country to another, usually with conditions, although the timescale is not as long as for multilateral aid.
Tied aid	Aid given with conditions, such as the receiving country having to buy manufactured goods from the donor country, or with political conditions.

Sample Questions *and Answers*

Question: General level 2000

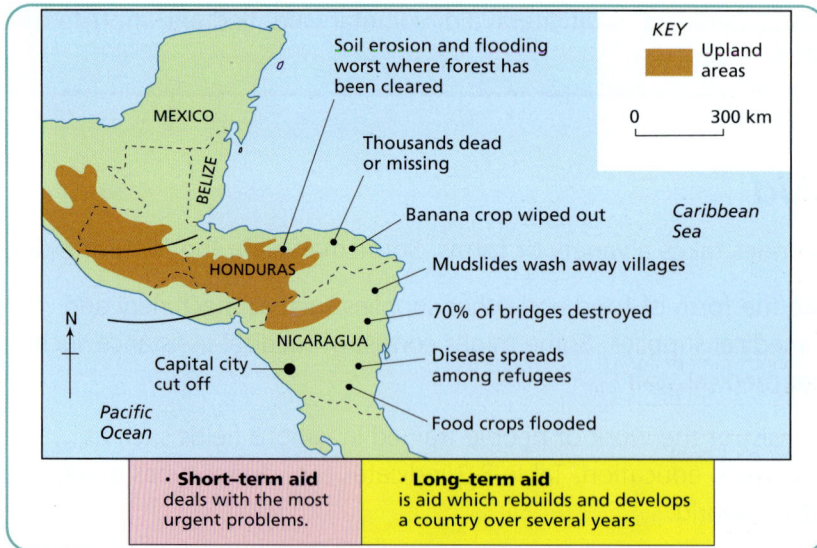

Figure 3.18 Hurricane disaster in Central America

Look at Figure 3.18.

'Only short-term aid can help Central America recover from Hurricane Mitch.' UN spokesman, 1998

Sample Questions and Answers continued

Do you agree with this statement?   Tick your choice   Yes ☐   No ☐

Give reasons for your choice.   **(4 marks ES)**

Sample answer

In the short term they need to keep alive. But in the long term the people will want to rebuild their country and rebuild their schools etc. (✔). They will also want clean water which comes from wells which are a long-term thing (✔). The most important things take time to build.

Comments and marks obtained

The answer has not made full use of the resource provided. It might have been easier to have discussed short-term aid in this situation. Having made a difficult choice the candidate found it difficult to find the correct things to say in the answer, ignoring the obvious benefits of short-term aid in a disaster situation. Only two valid points were made about the advantages of long-term aid, namely education and wells, gaining **2 marks from a possible 4**.

Question: Credit level 1999

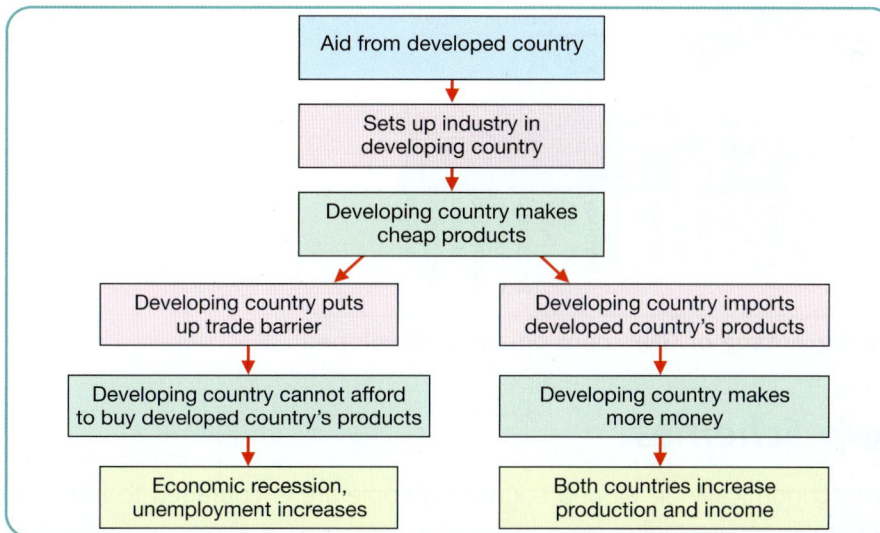

Figure 3.19 Possible effects of giving aid to developing countries

Look at Figures 3.19 and 3.20.

'Developed countries are keen to give aid to developing countries because of the benefits involved.' Do you agree with this statement? Give reasons for your answer.   **(4 marks ES)**

Sample Questions and Answers continued

Sample answer

It can't be that great a benefit if over half the countries listed don't give the UN target figure (✓). However, they are all giving some kind of aid so it can't be that bad. I think most countries think it is a good idea as all give something (✓). It must just be that governments don't want to waste all their money in case an economic recession started and they get affected. I think that they don't want to throw money away easily but think it looks a good plan on the whole (✓).

Comments and marks obtained

The answer is not very well written but some concrete points are made including reference to half of the donating countries not achieving the UN target and countries giving at least some form of aid. The notion that governments are cautious about the amount of money given in case their own economies might suffer is also worth a mark. In total the answer has obtained **3 marks out of a possible 4**.

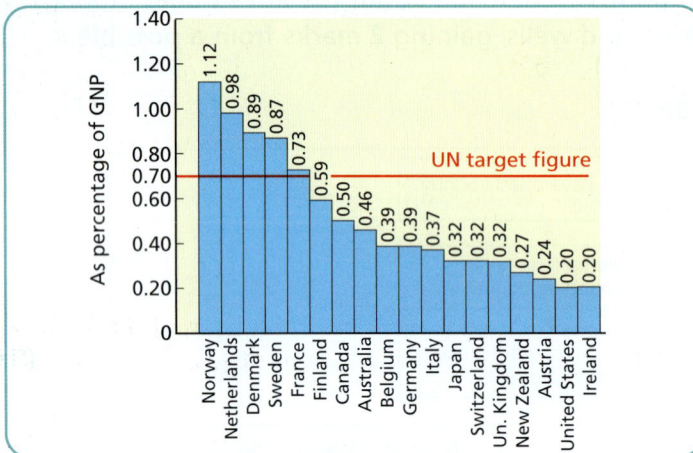

Figure 3.20 Percentage of GNP spent on aid by developed countries

Self-help Schemes

In many less economically developed countries people are encouraged to introduce self-help schemes, for example:

◆ new farming methods involving, such as 'miracle seeds' and soil conservation methods

◆ low or medium-level technology schemes in industry

◆ the use of basic irrigation schemes to improve local water supplies

◆ housing improvement schemes in shanty towns and primary health care schemes.

There are different levels of technology which can be used in all of the above. Figure 3.21 below shows some examples of these various levels of low/medium/intermediate and high-technology methods. Each of these has different advantages. The lowest types have the advantage of being the cheapest and the most readily available. The higher-level systems may increase the cost but may also bring more benefits to the population.

Sample Questions *and* Answers

Question: General level 2002

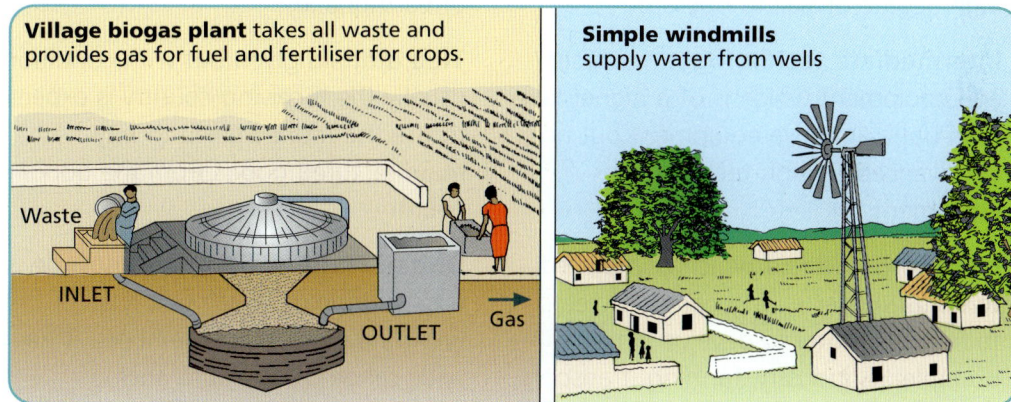

Village biogas plant takes all waste and provides gas for fuel and fertiliser for crops.

Waste

INLET

OUTLET   Gas

Simple windmills supply water from wells

Figure 3.21 Examples of intermediate technology

Look at Figure 3.21.

Explain why intermediate technology, such as that shown in Figure 3.21, is suitable for many rural communities in developing countries. **(4 marks KU)**

Sample answer

The technology is not very expensive (✓) so people from developing countries will not find themselves with huge debts (✓). The technology is simple so little training is required (✓). The village biogas plant gives the village a source of energy (gas) for free (✓).

Comments and marks obtained

The answer makes four very good points on the advantage of having this kind of technology by referring to cost, avoiding debts, no need for training and freely available energy source.

This merits the full **4 marks out of 4**.

Glossary of Associated Terms
Key Idea 17: Aid and Self-help Schemes

Appropriate technology: the use of machinery and equipment which is best suited to the needs, skills and wealth of local communities.

Bilateral aid: aid given from one country to another.

High technology: the use of advanced, sophisticated machinery which requires a high degree of skill to operate.

Intermediate technology: machinery and equipment which is of a higher level than basic primitive equipment but not as advanced as high-technology equipment.

Long-term aid: aid which is intended to be used over a long period of years to help a country to develop, for example, its industry, farming, transport system, education and health care systems.

Low technology: equipment which is very basic and cheap such as ox-drawn ploughs or wood-burning ovens.

Multilateral aid: aid given by a group of countries to poorer countries through agencies such as the United Nations.

Official aid: bilateral or multilateral aid given to a country.

Self-help schemes: projects in which local people become involved to improve their own living conditions.

Short-term aid: aid given immediately to help an area recover from a major disaster such as a flood, drought, famine or earthquake.

Tied aid: aid given but with conditions – usually the receiving country is expected to use the money given to buy manufactured goods from the donor country.

United Nations (UN): a worldwide organisation of which nearly all of the world's countries are members. It has a wide range of functions such as consideration for world health and world finance. It provides peacekeeping forces and includes various aid agencies.

Voluntary aid: aid given through charitable organisations such as the Red Cross and Oxfam.